NECHES RIVER
USER GUIDE

RIVER BOOKS

Sponsored by the

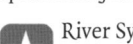 River Systems Institute
at Texas State University

Andrew Sansom, General Editor

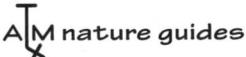 nature guides

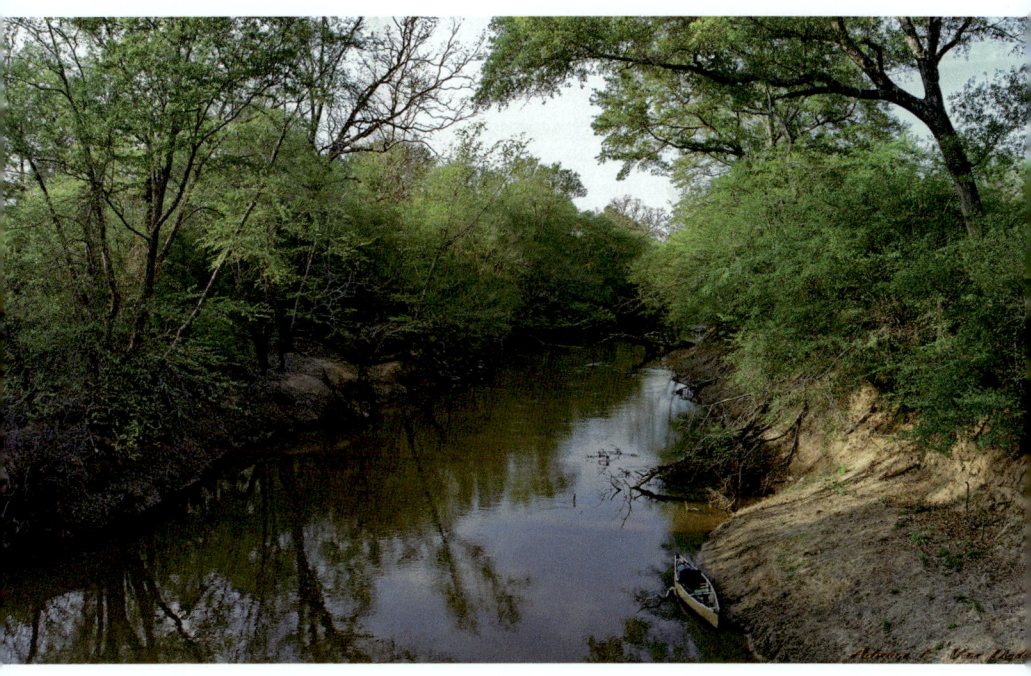

Neches River User Guide

By Gina Donovan

With Stephen D. Lange & Adrian F. Van Dellen

TEXAS A&M UNIVERSITY PRESS   COLLEGE STATION

Library of Congress
Cataloging-in-Publication Data

Donovan, Gina, 1965–
    Neches River user guide / by Gina Donovan ;
with Stephen D. Lange and Adrian F. Van
Dellen — 1st ed.
        p. cm. — (River books) (ATM nature
guides)
        ISBN-13: 978-1-60344-138-4 (waterproof
pbk. : alk. paper)
        ISBN-10: 1-60344-138-7 (waterproof pbk. :
alk. paper)
        1. Neches River (Tex.)—Guidebooks.
2. Neches River Watershed (Tex.)—
Guidebooks.   3. Stream ecology—Texas—
Neches River.   I. Lange, Stephen D.
II. Van Dellen, Adrian F.   III. Title.
IV. Series: River books (Series)
V. Series: TAM nature guides.
F392.N35D65   2009
917.64'1502—dc22
2009018525

Liability Waiver: The *Neches River User Guide* is published for the purpose of offering ONLY suggestions for the best possible ways to access and enjoy the Neches River. Due to the possibility of personal error, typographical error, misinterpretation of information, and the many hazards, both natural and man-made, the *Neches River User Guide*, its owners, employees, and all other persons or organizations, directly or indirectly associated with this publication, assume no responsibility for accidents, injury, death, or damage incurred by individuals or groups using this publication.

Neches River User Guide

A project of Texas Conservation Alliance

Creator and Editor: Gina Donovan
GIS Specialist: Stephen D. Lange
River Consultant: Adrian F. Van Dellen

Contributing Editors: Janice Bezanson, Brandt Mannchen, Julie Shackelford
Archeology Contributor: John Ippolito
History Contributors: Richard Donovan, Jonathan Gerland
Flora Contributors: Gina Donovan, Greg Grant, Elyce Rodewald, Larry Shelton
Fauna Contributors: Gary Calkins, Gina Donovan
Bird Consultants: Susan Billetdeaux, Winnie Burkett, Gina Donovan, Flo Hannah, Cliff Shackelford

Royalties from the sale of this user guide will go toward conserving and protecting the Neches River for future generations.

Texas Conservation Alliance extends its deep appreciation to the Houston Audubon Society, Stephen F. Austin State University Pineywoods Native Plant Center, Texas State University River Systems Institute, TLL Temple Foundation, Texas Parks and Wildlife Department, The Conservation Fund–Texas, and the USDA Forest Service for their assistance in developing this guide.

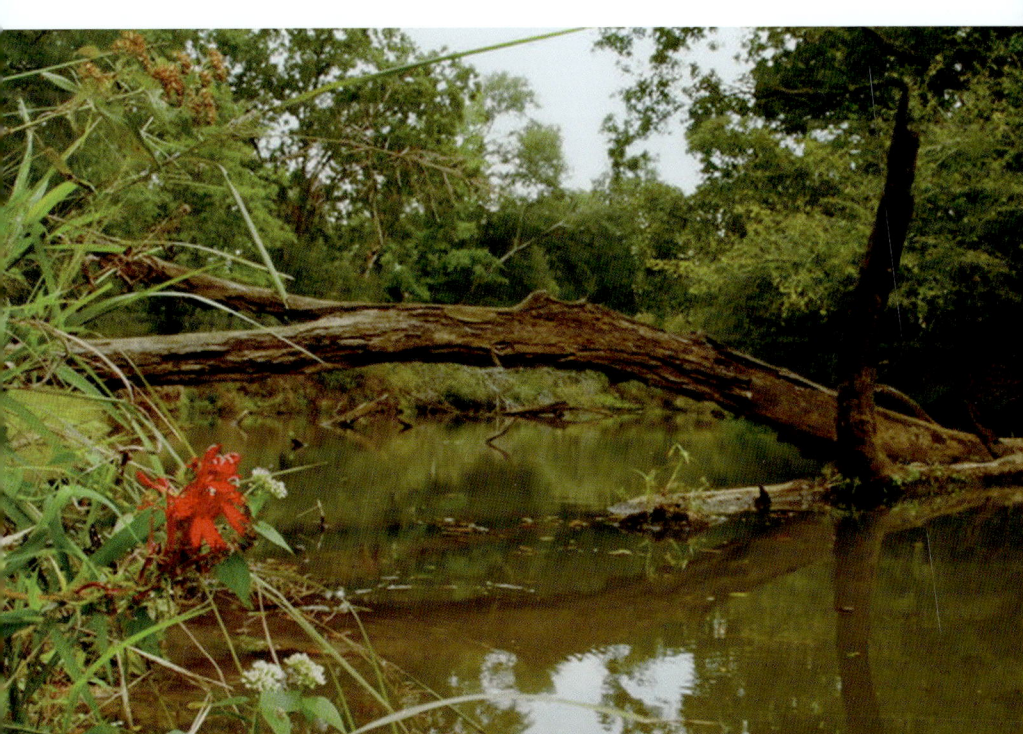

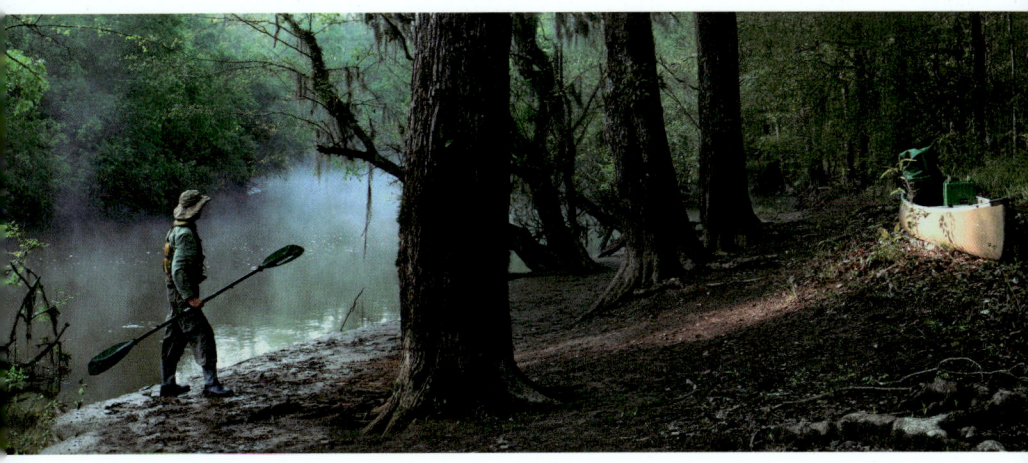

Contents

SERIES EDITOR'S FOREWORD

On the day that I write this, we spread the ashes of conservationist Ned Fritz in a garden in Dallas. During the last half of the twentieth century, Ned was the foremost champion of the Neches River and its environs. He knew that one of the best ways to really get to know a place is when a river runs through it. Nowhere in Texas is this more true than along the Neches, which meanders through some of the most biologically diverse forests, swamps, and savannahs in America. With this latest volume of River Books, Gina Donovan equips readers to make their own memorable journeys of natural and cultural exploration and adventure along the river.

In the *Neches River User Guide*, you will find a practical introduction to the history and archeology of the region, a broad explanation of the river's natural origins, and a helpful field guide to its flora and fauna. Additionally, Gina Donovan, now executive director of Houston Audubon and a member of one of East Texas' most distinguished conservation families, has produced the first comprehensive river map for traversing the waterway. The maps and descriptions give clear instructions as to where to put in and take out, as well as a section-by-section description of what the paddler will find around each bend.

This unique guide would not have been possible without continued support from the TLL Temple Foundation, which also funded the very first volume of River Books, *Paddling the Wild Neches*, which gives the account of an actual voyage down the river by Gina's father, Richard Donovan. The Foundation, created with resources of one of Texas' most prominent business families, has long been a leading force for conservation and ecotourism in East Texas and beyond. Its support of this guide is emblematic of a fundamental understanding of the relationship between responsible recreational use of our natural resources—including hunting, fishing, camping, hiking, bird watching, and paddling—and preservation and conservation of those resources. Generally, people who challenge themselves to enjoy splendid places like the Neches River Watershed also constitute the principal defenders and protectors of such special places.

People like Ned Fritz taught many of us that, on a lovely river like the Neches, it is possible to leave everything behind but the river itself and become fully immersed in the resource while at the same time developing a profound sense of responsibility for protecting it for future generations.

So, as you pack this little guide in your gear and head down the Neches, celebrate its beauty and diversity and don't take it for granted. If you can, take a child with you. It will be a beneficial experience for both of you, an experience that may never be forgotten. And, it will be good for the future of the river.

—Andrew Sansom

Neches River
User Guide

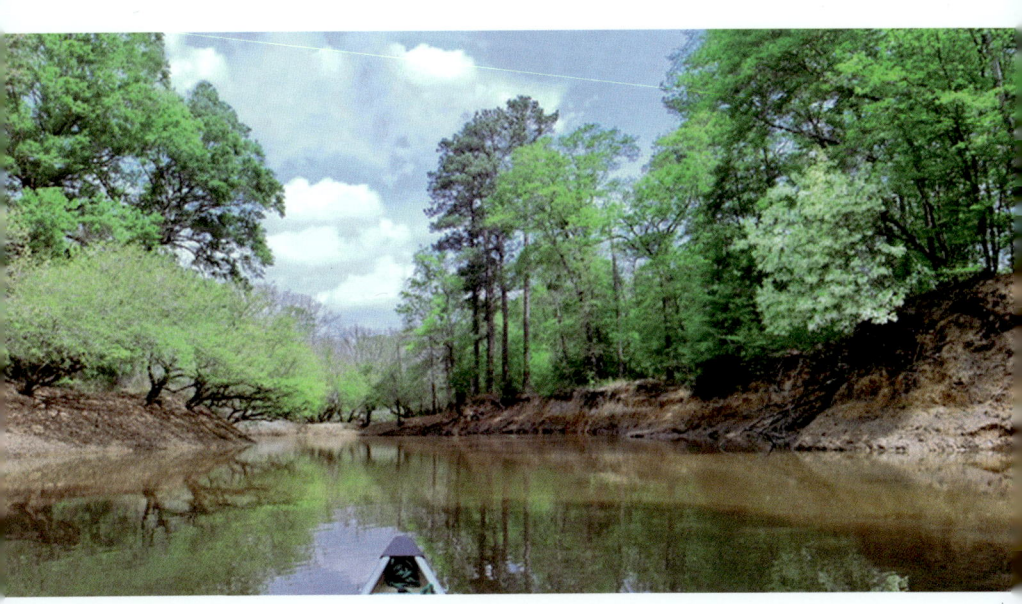

Introduction

The soul-stirring hoots of the barred owl. . . the lonesome sounds of the coyote. . . or the gentle riffles of a canoe sliding through the tannin-colored water. . . A trip down the Neches River in East Texas always holds the promise of adventure. The Neches is truly one of Texas' wildest and least-discovered natural assets. The *Neches River User Guide* is intended to make this treasure a little more accessible to those interested in unlocking the river's secrets. With some advanced planning, the reader can enjoy a river trip ranging from an afternoon float to a several-week expedition. The guide's brief descriptions about the river's archeology, history, and wildlife provide a launching point for the reader to begin the singular Texas experience that is the Neches River.

Rising just east of Colfax in eastern Van Zandt County, the Neches River flows more than 400 miles into Sabine Lake, a fertile estuary near Port Arthur, and ultimately empties into the Gulf of Mexico. White sandbars frequently dot the edges of the Neches River's mocha-colored waters, disappearing only during periods of high floodwaters. The river's freshwater flood flows nourish the forests of the Neches

River National Wildlife Refuge, the Davy Crockett and Angelina national forests, Big Slough and Upland Island wilderness areas, the Pineywoods Mitigation Bank, the Alabama Creek, and Angelina, Neches, and Dam B wildlife management areas, the internationally recognized Big Thicket National Preserve, and thousands of acres of privately owned lands.

Two reservoirs are located on the Neches—Lake Palestine, southeast of Tyler, and Lake B. A. Steinhagen near Town Bluff, west of Jasper.

The tree- and shrub-covered banks of the Neches and its slow-moving current provide a serene setting for canoeists, anglers, hunters, birders, and wildlife watchers. In its upper reaches, the river traverses rolling terrain. Its sometimes steep, heavily wooded banks and bluffs grow a combination of pines and hardwoods—oak, hickory, dogwood, hackberry, pecan, and blackgum. In its lower reaches, the Neches flows through a generally flat landscape, and the vegetation consists largely of water-tolerant hardwoods mixed with pine trees and a diverse range of shrubs and grasses. On the far southern end, cypress and tupelo abound, especially

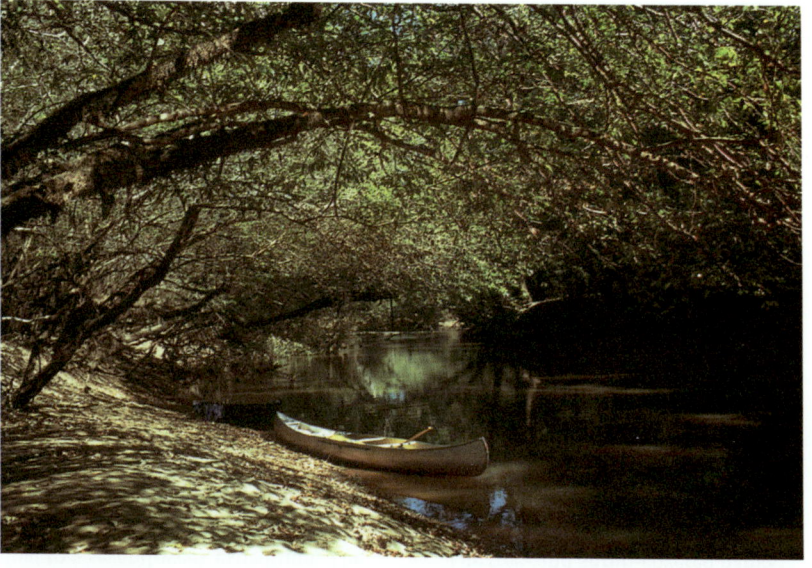

Planer Trees over-arch canoes. Courtesy Gina Donovan

in the meanders and bayous that drain into the Neches just north of Beaumont. High rainfall rates produce frequent flooding of low-lying areas, increasing the diversity of wildlife habitat, especially in the Big Thicket National Preserve.

The diversity and productivity of these forested wetlands provide refuge and food sources for many mammals and migrating songbirds, waterbirds, and woodland waterfowl. The Neches River's bottomlands are rich with squirrels, raccoons, deer, coyotes, armadillos, opossums, rabbits, and a variety of reptiles and amphibians. Bottomland forests are important for many rare breeding birds, such as the swallow-tailed kite, American woodcock, and

prothonotary and Swainson's warblers, and are absolute necessities for land birds during migration, especially spring migration. The river supports threatened and endangered species, including the bald eagle, peregrine falcon, wood stork, American swallow-tailed kite, Louisiana black bear, Rafinesque's big-eared bat, paddlefish, and alligator snapping turtle.

The Neches River Valley is rich not only with abundant trees, vegetation, wildlife, and birds but also with culture, history, and recreational opportunities. The following sections of this guide will give a brief synopsis of life along the Neches from primitive times to the present.

ARCHEOLOGY

The Neches River has been an essential element of the cultures in East Texas throughout human habitation in the region. From the earliest big-game hunters to the coming of Europeans, the river has been the principal means of transportation and communication. Its fertile bottomlands provided a veritable supermarket of resources for prehistoric inhabitants.

Clovis and other early spear points found at various sites along the central Neches Valley attest to the presence of Paleo-Indian big-game hunters in the region. Hunter-gatherers were predominant for thousands of years. In the last several centuries B.C., subtle changes started taking place—simply crafted utilitarian pottery began to appear, indicating a shift toward a more sedentary and settled lifestyle. The beginning of crop agriculture appeared about this time all across the southeastern United States. As agriculture succeeded, populations grew in size and complexity, and these woodland peoples developed politically and religiously stratified societies. They developed large ceremonial centers, dominated by high mounds supporting the holiest of their temples.

In the forest reaches of the Neches River Valley, the woodland cultural characteristics were ascribed to an American Indian tribe known as the Caddo Nation. Sites attributed to the Caddo Indians are present all along the middle Neches River and its tributaries. Their culture flourished and grew for more than a thousand years. When French and Spanish explorers came to East Texas in the sixteenth century, it was the Caddo they first encountered.

Along the lower Neches, after the river leaves the pine forests and enters the open coastal prairies and marshes, the hunting and gathering subsistence strategies continued unabated until the eighteenth century. Various tribes, whose names have been lost in the mists of time, were experts at exploiting the resources found in the coastal prairie borderlands of southeast Texas. Mounds of oyster and clam shells mark the settlements of these people all along the upper Texas coast. Their basic utilitarian pottery, developed sometime before the birth of Christ, persisted virtually unchanged until contact with European explorers and settlers in the sixteenth and seventeenth centuries.

Once the first Anglos were established along the Neches Valley, settlement expanded rapidly. By the early nineteenth century, most of the American Indian inhabitants had left the region. Some had left voluntarily, while others had been forcibly removed to reservations in Oklahoma, exterminated, or absorbed into other tribal entities. Many others had died of introduced diseases. So, European newcomers faced little competition for land or resources. The age of Texan expansion had dawned.

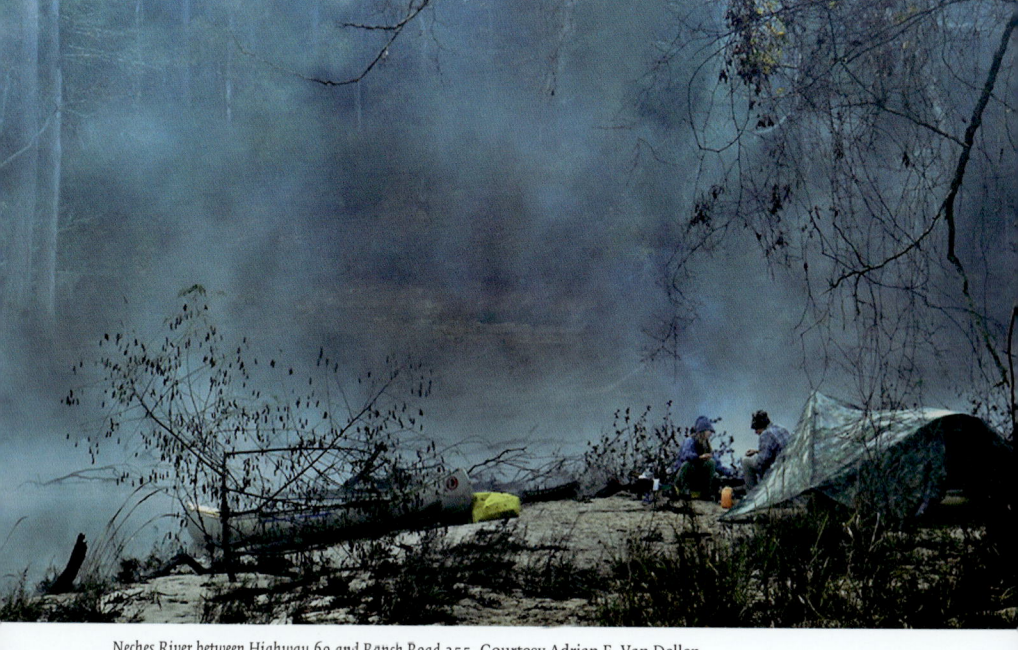

Neches River between Highway 69 and Ranch Road 255. Courtesy Adrian F. Van Dellen

HISTORY

In a world of superhighways, parking lots, and skyscrapers, the Neches stands alone as East Texas' last remaining wild river. The towering forests along its banks and the river's meandering current look much as they did in 1820 when Moses Austin, Stephen F. Austin's father, forded his horse across the river on Texas' first great thoroughfare, called El Camino Real (or, the King's Highway). Today, Texas Highway 21 closely follows this historic route, which was used by American Indians and buffalos hundreds of years before Domingo Terán de los Ríos blazed the route in 1691.

The Spanish gave the Neches its current name, as they did most of the rivers and sizeable streams of the state. The Caddo Indians, who had inhabited the Neches Valley since about 800 A.D., called the river Nachawi, their name for the bois d'arc trees whose wood they used to make their hunting bows. The first European settlement built in Texas, mission San Francisco de los Tejas, was built by the Spanish in May of 1690 near where San Pedro Creek joins the Neches. Today Mission Tejas State Park is near the spot where Spanish friars toiled to grow crops and convert Indians to Christianity more than three hundred years ago.

Anglos began settling along the Neches two decades before Stephen F. Austin established his first colony on the Brazos. The Texas forest country was a near paradise for these early pioneers. Trees from the forest were readily available for building homes. Natural foods were plentiful—wild game (deer, ducks, squirrels, pigeons, black bear, cougars, and feral hogs) was abundant, many different fruits and nuts ripened seasonally, and the river was teeming with fish. One early traveler wrote that when the river was low, it appeared he might have been able to walk across on the backs of fish! One old settler lamented he had difficulty hearing the tinkle of his cowbell over the gobbling of wild turkeys. Another said he had trouble deciding which deer to kill from his front porch when the family needed fresh meat.

As more and more people from the United States swam their horses, cattle, and hogs and floated their wagons across the Sabine and Red rivers into the province of Tejas, the Mexican government became concerned and dispatched General Manuel de Mier y Terán to curtail the flow. One of Terán's first acts was to construct a fort at a much-used river crossing on the Neches. Long known as the "Pass to the South," the natural, hard rock bottom crossing had been easily used by buffalo, other wildlife, and Indians for hundreds of years. The fort proved unsuccessful. A granite marker erected on a high rock-strewn hill overlooking the Neches a short distance upriver from the mouth of Shawnee Creek is all that remains of Fort Teran (see page 36). In the years since its abandonment during the Texas Revolution in 1836, rumors have persisted of buried treasure either in the floor of the fort's supply cave or submerged in the river.

The steady flow of Anglo settlers bred a population of "Texians" fretting for independence from Mexico. Sam Houston negotiated a treaty with the few remaining Caddos and the numerous other tribes, including Cherokee, Shawnee, and Choctaw, that had been forced into Texas from the United States. The treaty granted these American Indians land north of the El Camino Real if they would remain neutral in the looming war with Mexico.

Sam Houston's stunning defeat of Mexican General Santa Anna at San Jacinto made this treaty of little value to the new Republic of Texas. In his first message to the Texas Congress, the second president of the republic, Mirabeau B. Lamar, called for "the prosecution of an exterminating war" on Texas Indians and further stated this war would "admit no compromise and have no termination except their total extinction or expulsion." He specifically addressed the Cherokees, saying they held no legitimate title to the land even though they had signed a treaty with Houston and had lived there for almost forty years, four times longer than most Anglos.

The rescinding of this treaty and the steady encroachment of settlers into Indian lands precipitated the infamous Killough Massacre in northern Cherokee County. On October 5, 1838, a band of Indians attacked the Woods, Killough, and Williams families as they were harvesting their crops. Eighteen members of those families were either killed or carried off. Those carried off were never heard from again.

In July of 1839, Lamar ordered Thomas J. Rusk and an army of five hundred men to move the Indians from Texas to Oklahoma Territory. The battle that ensued in Cherokee, Henderson, and Van Zandt counties resulted in the death of many Indians. Many more died of cold and hunger in Oklahoma. When the body of Cherokee Chief Bowles was found on the battlefield, he was wearing a sword presented to him by Sam Houston.

With the defeat of the Mexicans and removal of the Indians, the trickle of U.S. settlers into Texas became a torrent.

These first arrivals were often "running from something" and quietly disappeared into the forested wilderness. They brought with them a way of life that endured almost unchanged in the Texas forest country until the 1960s—a hunter-gatherer, farmer-stockman society. Early settlers raised corn, squash, beans, sweet potatoes, and melons. They cared little for owning the land; they used it as a "commons," much as the Indians had done. The settlers also hunted native wildlife, and turned nearly wild hogs and cattle loose in the bottoms, to be rounded up at killing time. They practiced a southern, forest-adapted tradition of stock raising that was vastly different from that found farther west. Using a plaited rawhide stock whip with a buckskin "cracker" tip to work cattle in thick woods, the stockman relied on his horse to keep him safe from his dangerous livestock. However, when an animal needed to be caught and held for castration, marking, or "doctoring," the stockman turned, not to his horse and rope, but to his dogs. These intelligent, aggressive, well-trained dogs were as valuable as any possession the stockmen owned.

In the early 1800s, the Neches was a major artery of commerce. By 1840 steamboats like the *Angelina* were plying the Neches waters and had opened the East Texas interior to a booming trade with New Orleans and Galveston. The riverboats that navigated the Neches were small and of shallow draft to avoid obstructions that could tear the bottom out of any boat. Boat captains particularly dreaded Rocky Shoals, near Rockland, and few ventured north of the shoals after the heavy spring rains had passed. (See photo on page 10.)

Steamboats transported passengers and freight up and down both the Ange-

lina and Neches rivers with decks and holds piled high with otter, beaver, mink pelts, and the skins of deer, bear, panther, and maybe buffalo. There would also be barrel staves, handmade baskets, gourds, and stone crocks filled with ribbon cane syrup, wild honey, and bear grease. As time progressed, the amount of cotton hauled downriver rapidly increased. The steamers' upriver cargos consisted of axes, saws, rifles, shotguns, ammunition, cloth, coffee, sugar, flour, salt, and the whole list of staples needed for living on the frontier of what was then called the "Texas Western Woods." (See photo on page 42)

The steamboat's dominance of transportation was brief. Railroads began to siphon off commerce by the 1880s. The last of the steamers, the Neches Belle, made its final run in 1894.

The paddle-wheelers had other serious problems besides competition from the railroads. Beginning barely five years after Texas' statehood in 1845, during every rise in the river, timber men would unleash thousands of pine and cypress logs into the bloated, swift current bound for Beaumont's new sawmills. Cutting trees anywhere with easy access to the river, these loggers ushered in the era of big timber. The floating logs clogged the river, and steamboats could not move for fear of puncturing their hulls. In every battle between boats and logs, the steamboats lost.

The Neches and its surrounding lands remained essentially a wilderness

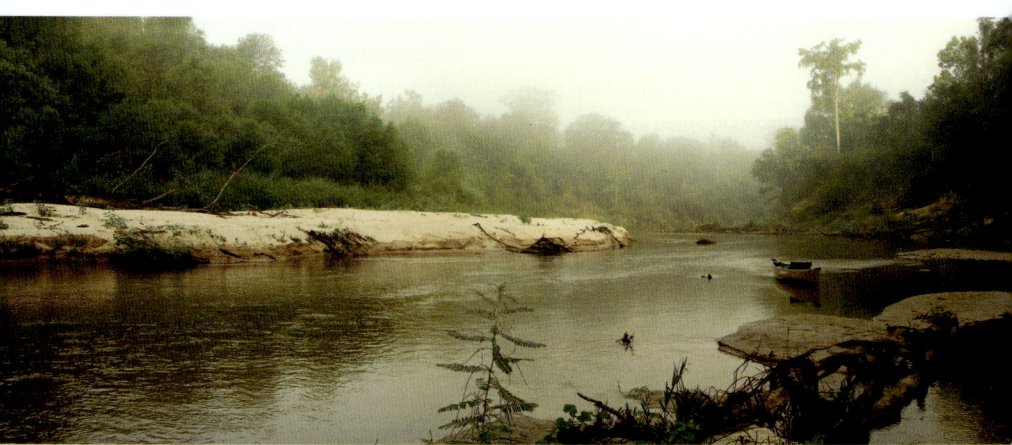

Sandbar at Billam's Creek. Courtesy Adrian F. Van Dellen

until the late nineteenth century when the coming of lumber mills wiped out Texas' virgin forests. These big mills are too numerous to mention by name, but most of them resembled the mill in the ghost town of Old Aldridge. The skeletal remains of this once-thriving sawmill can be seen a few hundred yards from the Neches River in Angelina National Forest, near Boykin Springs Campground.

Up and down the Neches, settlements established at many of the river crossings eventually grew into cities and towns. This brought change—not planned or orderly change but change born of inescapable forces brought about by different life circumstances. Changing

job markets drew people from the rural Neches country to better paying jobs in the cities. Timber markets evolved, and timber companies' production strategies changed. The river and its surrounding forests were no longer required to provide the essentials of life for families that had lived along its margins for generations. Pressure on wildlife from hunting, fishing, and trapping was reduced to a fraction of what it had been, allowing the resurgence of wildlife visible along the Neches today. Portions of the forests have been allowed to mature and offer the majestic beauty and plant variety found in few, if any, other places in North America today.

KNOW BEFORE YOU GO

Public access to the Neches River is usually via public property. Typical access may be from the right-of-way of a public road that crosses the river, at a publicly owned boat launch area, or from some other publicly owned land (a local or state park or national forest, for example) adjacent to the river. There is no general right to cross private property to access the river or for camping or hunting. Please do not trespass on private lands. Use of private land adjacent to the river without landowner permission can be considered trespassing. Under Texas Penal Code (30.05), criminal trespass occurs when one enters property after receiving notice not to enter. Notice includes verbal notice, a fence, sign(s), purple paint on posts or trees, or the visible presence of crops grown for human consumption.

Texas courts have adopted the "gradient boundary" as the usual dividing line between public ownership of a stream's bed and lower bank area, and private ownership of the higher bank area and uplands beyond. Thus, there is generally no question as to the public's right to use the bank area up to the gradient boundary. However, determining the boundary is a complex task performable only by specially trained persons. The mean gradient boundary is located midway between the lower level of the flowing water that just reaches the so-called "cut bank," and the higher level of the flowing water that just does not overtop the cut bank. The cut bank is located at the outer edge of a stream's bed, separating the bed from the adjacent upland and confining the waters to a definite channel. This information is found on the Texas Parks and Wildlife Department's Web site at http://www.tpwd.state.tx.us. For further explanation, please visit the "publications" section of the Web site or do an Internet search for "gradient boundary" or "stream navigation law."

Public boat ramps along the Neches River are scarce, but due to the smaller sizes of canoes and kayaks, these crafts can be launched at public access points fairly easily. Be aware that terrain at these points can be steep, muddy, and located on busy highways. Always use caution when launching and retrieving watercraft.

Parking at public boat ramps and along highways is at your own risk. Generally, automobiles remain unharmed, but you may wish to have someone drop you off and pick you up at specified locations to avoid the risk.

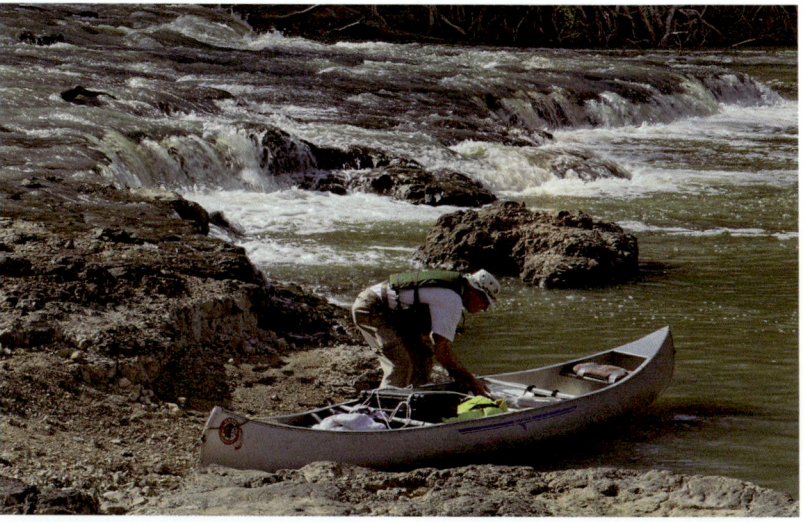

Canoer at Rocky Shoals waterfall. (see page 38)
Courtesy Adrian F. Van Dellen

Due to numerous logjams on the upper portions of the Neches River (north of U.S. Highway 69) and the need to portage over, around, or through the majority of them, paddlers need to be in good physical condition. Please note your physical condition and prepare accordingly.

Drinking water, food, sunscreen, insect repellant, proper footwear (water shoes or tennis shoes), personal flotation devices, sunglasses, and a hat are advised.

Cell phone service is rarely available. Do not rely on a cell phone in case of emergency. It is advisable to tell someone where you are going, the approximate time you plan to return, and who to contact in case of an emergency.

Numerous species of wildlife call the Neches River corridor home. Please note alligators and snakes can present unpleasant, even life-threatening situations. Do not try to approach or touch any wildlife, as wild animals can scratch or bite and cause serious injury.

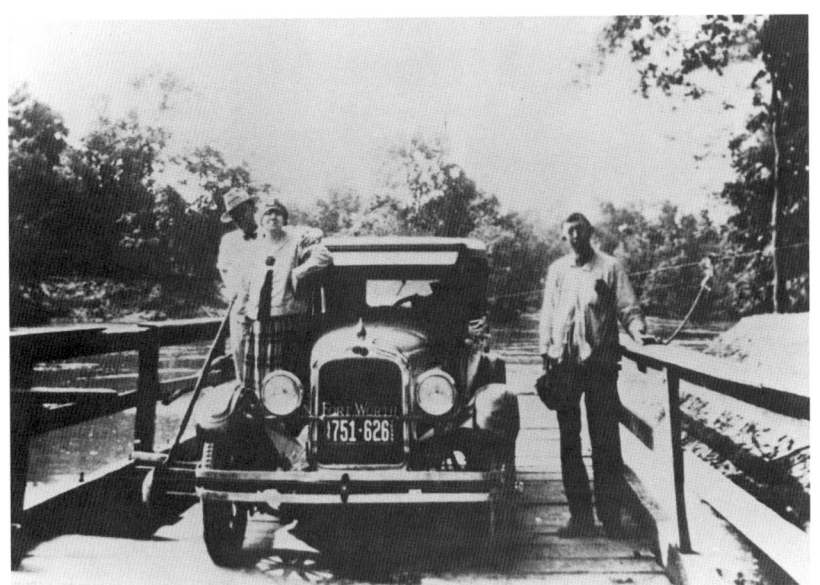

Bonnie and Clyde at Dunkin Ferry (see page 36). Courtesy Kenneth B. Hill

Neches River Watershed
of East Texas

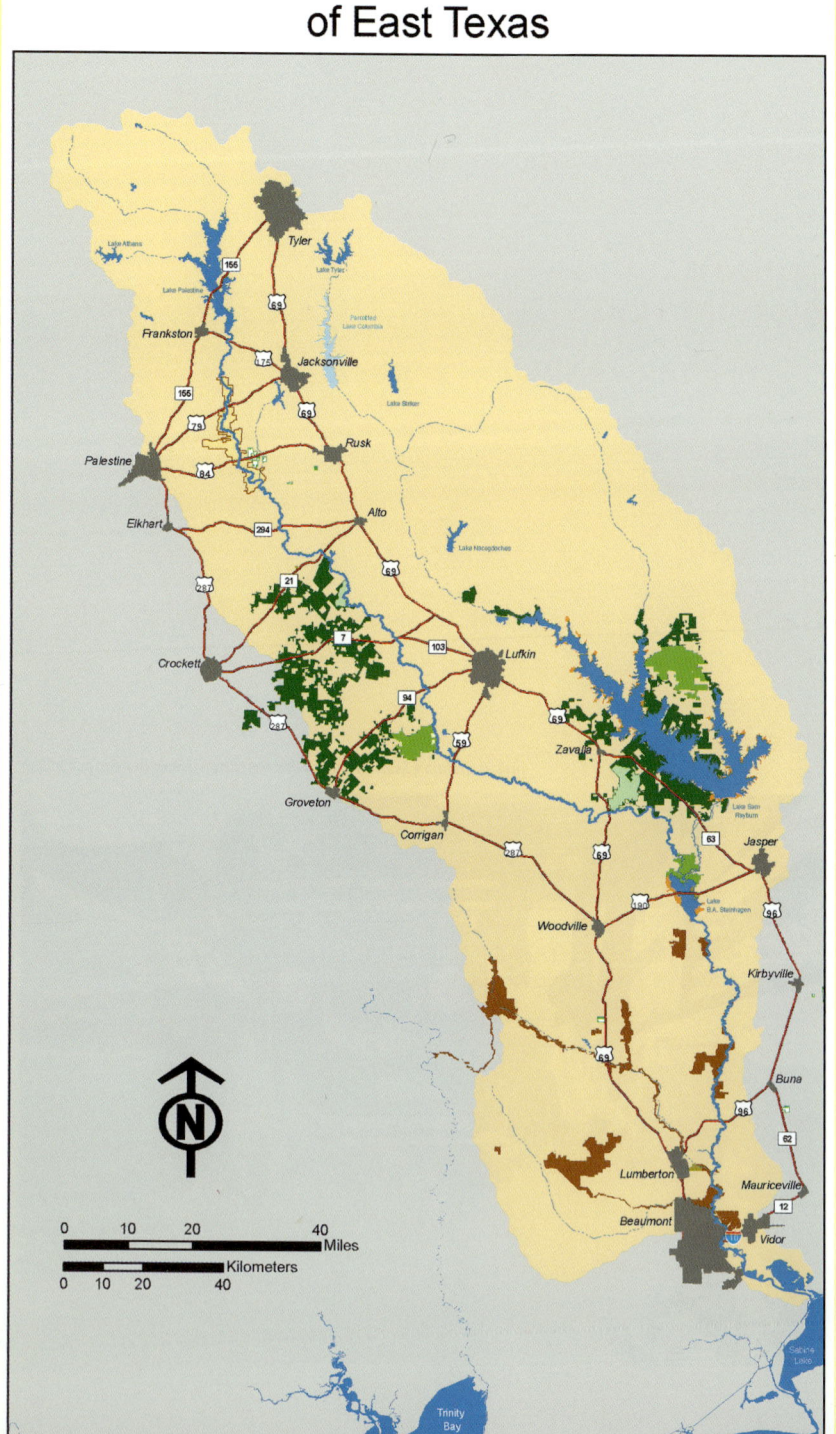

River Maps & Points of Interest

The following maps divide the Neches River into eighteen navigable segments, with side notes to help users make the most of their experience. GPS coordinates will help locate points of interest along the way.

Legend for Maps

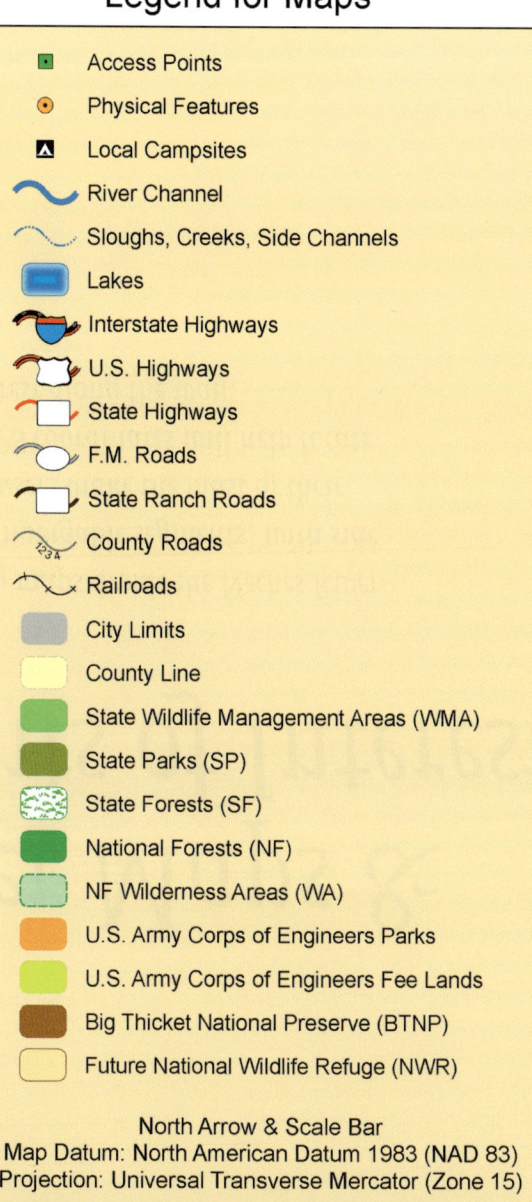

■ Access Points

⊙ Physical Features

▲ Local Campsites

〜 River Channel

〜 Sloughs, Creeks, Side Channels

▭ Lakes

〜 Interstate Highways

〜 U.S. Highways

〜 State Highways

〜 F.M. Roads

〜 State Ranch Roads

〜 County Roads

〜 Railroads

▨ City Limits

▢ County Line

▨ State Wildlife Management Areas (WMA)

▨ State Parks (SP)

▨ State Forests (SF)

▨ National Forests (NF)

▨ NF Wilderness Areas (WA)

▨ U.S. Army Corps of Engineers Parks

▨ U.S. Army Corps of Engineers Fee Lands

▨ Big Thicket National Preserve (BTNP)

▢ Future National Wildlife Refuge (NWR)

North Arrow & Scale Bar
Map Datum: North American Datum 1983 (NAD 83)
Projection: Universal Transverse Mercator (Zone 15)

0	0.5	1	2

Miles

0	0.5	1	2

Kilometers

Neches River
and Points of Interest

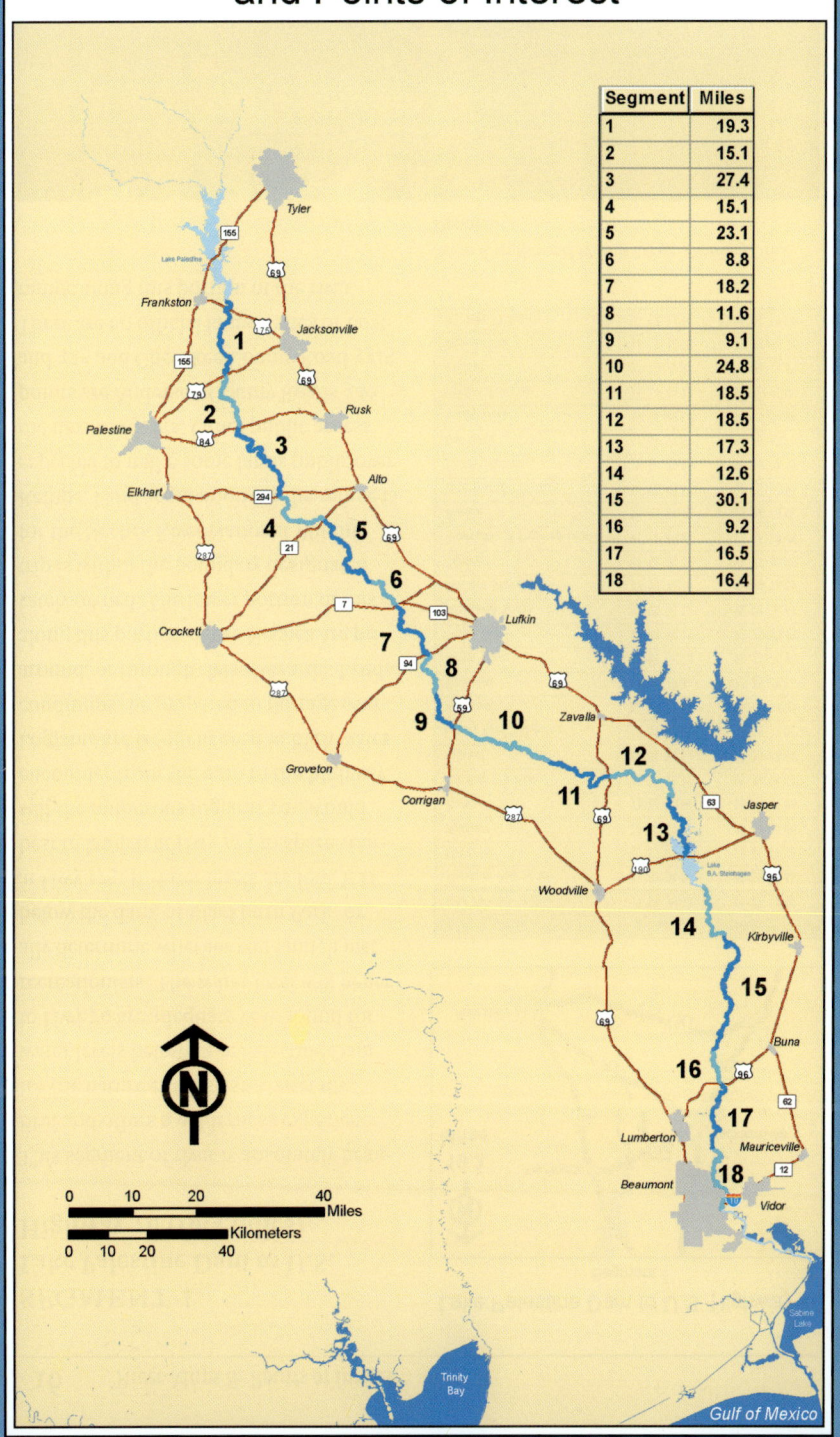

Segment	Miles
1	19.3
2	15.1
3	27.4
4	15.1
5	23.1
6	8.8
7	18.2
8	11.6
9	9.1
10	24.8
11	18.5
12	18.5
13	17.3
14	12.6
15	30.1
16	9.2
17	16.5
18	16.4

SEGMENT 1
Lake Palestine Dam to U.S. Highway 79 (19.3 miles)

This segment of river is absolutely beautiful and offers a wilderness experience for the nature enthusiast. Generally, water levels below Lake Palestine Dam to Hwy 79 are adequate year-round for recreationists. The water level will generally determine whether you launch just below the dam, at Blackburn Park, or at Hwy 175. If water levels are low, it is best to launch at Hwy 175 to spare yourself the numerous logjams you would encounter from the dam to the highway. Logjams are frequent even in high water conditions. Be prepared to portage over, around, or through these hazards. Lands along this portion of the Neches are privately owned. The lower portion of this trip is within the boundary designated for the Neches River National Wildlife Refuge. Land acquisition for the refuge is expected to make some lands public over the next few years. Other public access points are Anderson County Roads 320 and 335 and Cherokee County Road 3315. There are no official launch sites or boat ramps along this portion of the river.

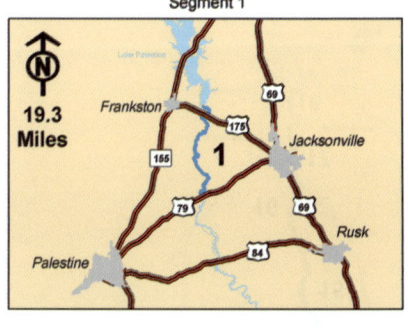

Lake Palestine Dam to U.S. Highway 79
Segment 1

Points of Interest	Latitude	Longitude
Lake Palestine Spillway	32 03 20.38	-95 25 55.30
Blackburn Park Access	32 03 04.81	-95 25 37.61
Flat Creek	32 02 49.00	-95 25 19.36
Pipeline	32 02 39.73	-95 25 21.40
Pipeline	32 02 20.68	-95 25 34.46
U.S. Hwy 175 Access	32 02 23.29	-95 26 08.36
Railroad	32 01 26.12	-95 25 40.66
County Road 320 Access	32 01 23.66	-95 25 34.92
Power line	32 01 21.32	-95 25 33.05
Power line	32 01 10.31	-95 25 28.96
Jordan Creek	32 01 08.70	-95 25 26.87
Caddo Creek	32 00 52.19	-95 25 40.43
Abandoned Bridge	32 00 13.12	-95 26 30.67
Kickapoo Branch Creek	31 59 22.48	-95 26 58.05
Pipeline	31 58 34.28	-95 27 08.10
Headwaters of Proposed Fastrill	31 57 50.67	-95 26 38.48
Pipeline	31 56 51.12	-95 26 48.05
County Road 335 Access	31 56 43.97	-95 26 40.92
Pipeline, Powerline	31 56 20.64	-95 26 16.89
Owl Creek	31 55 34.51	-95 25 12.70
Brushy Creek	31 54 49.51	-95 25 54.06
Pipeline, Powerline	31 54 44.27	-95 25 52.09
Railroad	31 53 52.87	-95 26 15.45
Walnut Creek	31 53 40.96	-95 26 06.73
U.S. Hwy 79 Access	31 53 33.92	-95 25 51.74

Neches River National Wildlife Refuge. Courtesy Gina Donovan

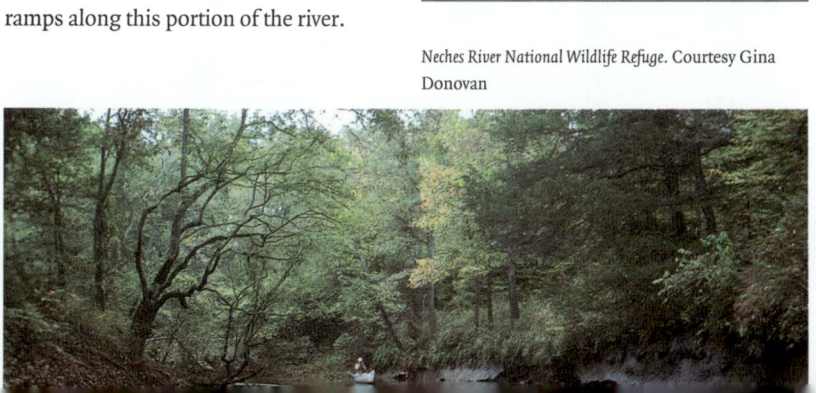

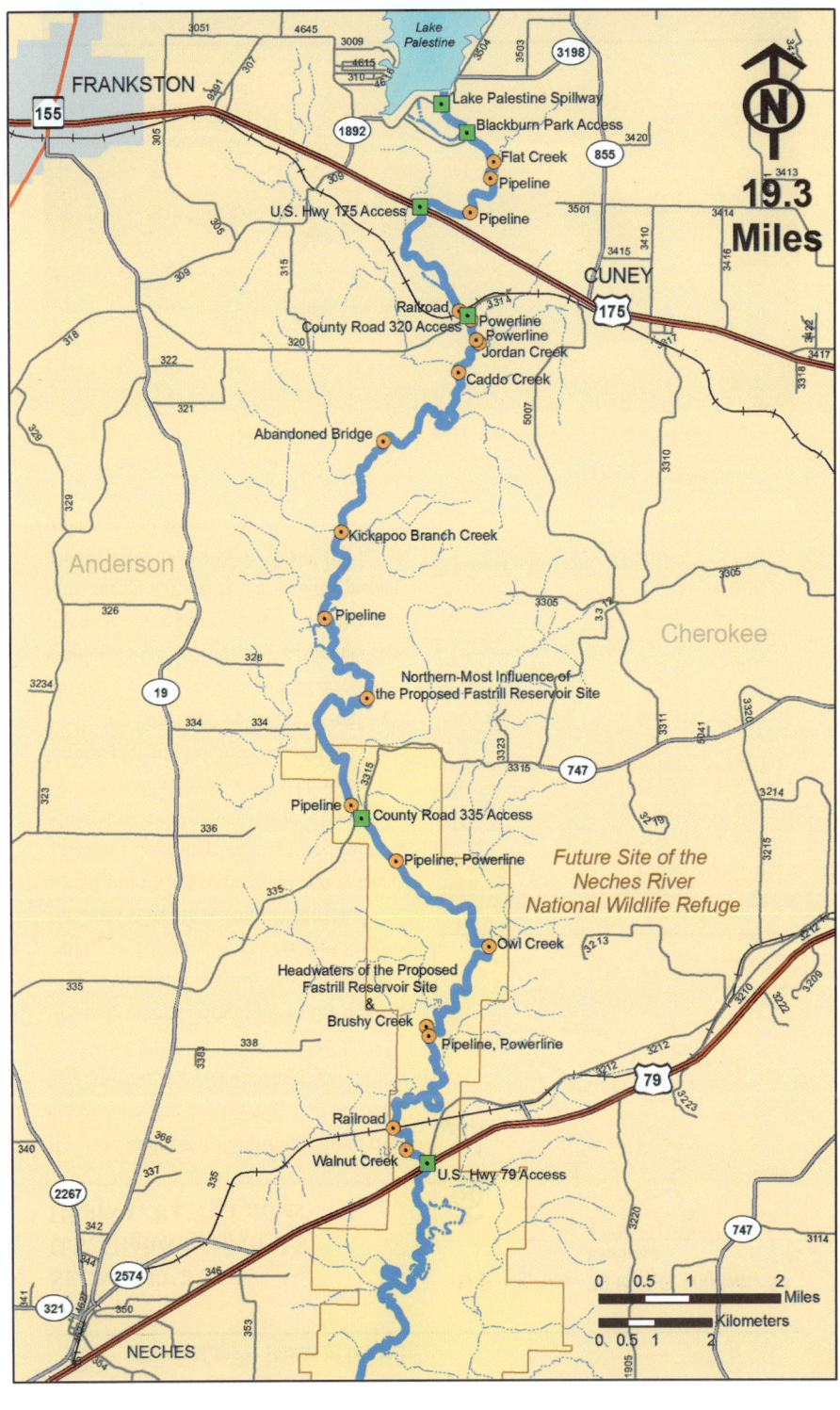

SEGMENT 2
U.S. Highway 79 to U.S. Highway 84 (15.1 miles)

If you're looking for another remote river experience, this is it. Again, use caution when portaging logjams, and remember lands along this segment of river are privately owned. Hopefully in the not too distant future, this entire river segment will be surrounded by the Neches River National Wildlife Refuge. Additional access points are Anderson County Road 4629 and Cherokee County Road 1906. There is a pseudo boat ramp at Hwy 84 with an asphalt road, but no concrete launch pad and no official parking.

Rocky Point *(31° 50′ 29.57″ / -95° 23′ 29.19″) A concrete weir constructed across the river here in the late 1960s by the City of Palestine is the intake point for the city's municipal water supply. CAUTION: The weir can provide a waterfall of up to 4 feet in height. It is a recommended portage.*

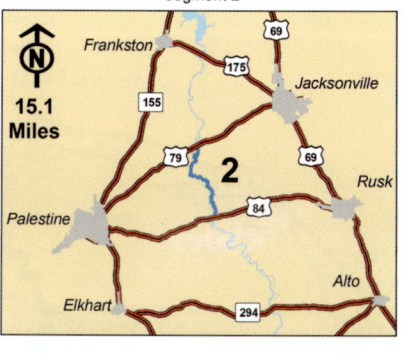

U.S. Highway 79 to U.S. Highway 84
Segment 2

Points of Interest	Latitude	Longitude
U.S. Hwy 79 Access	31 53 33.92	-95 25 51.74
Pipeline	31 53 24.27	-95 25 58.28
Pipeline	31 52 54.18	-95 25 55.92
White Oak Creek	31 51 25.46	-95 26 43.88
Rocky Point Spillway (Portage)	31 50 29.57	-95 26 29.19
County Road 354 Access	31 50 36.45	-95 26 17.40
Hurricane Creek	31 49 33.48	-95 24 44.53
Pipeline	31 49 10.93	-95 24 11.59
Old Channel	31 48 38.02	-95 24 15.22
Powerline	31 48 05.85	-95 24 11.58
Old Channel	31 47 01.06	-95 23 45.30
U.S. Hwy 84 Access	31 46 36.97	-95 23 46.73

Rocky Point concrete wier. Courtesy Richard M. Donovan

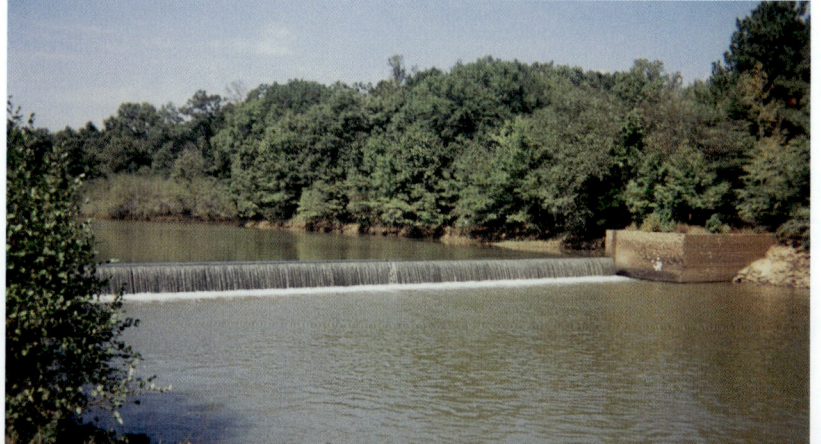

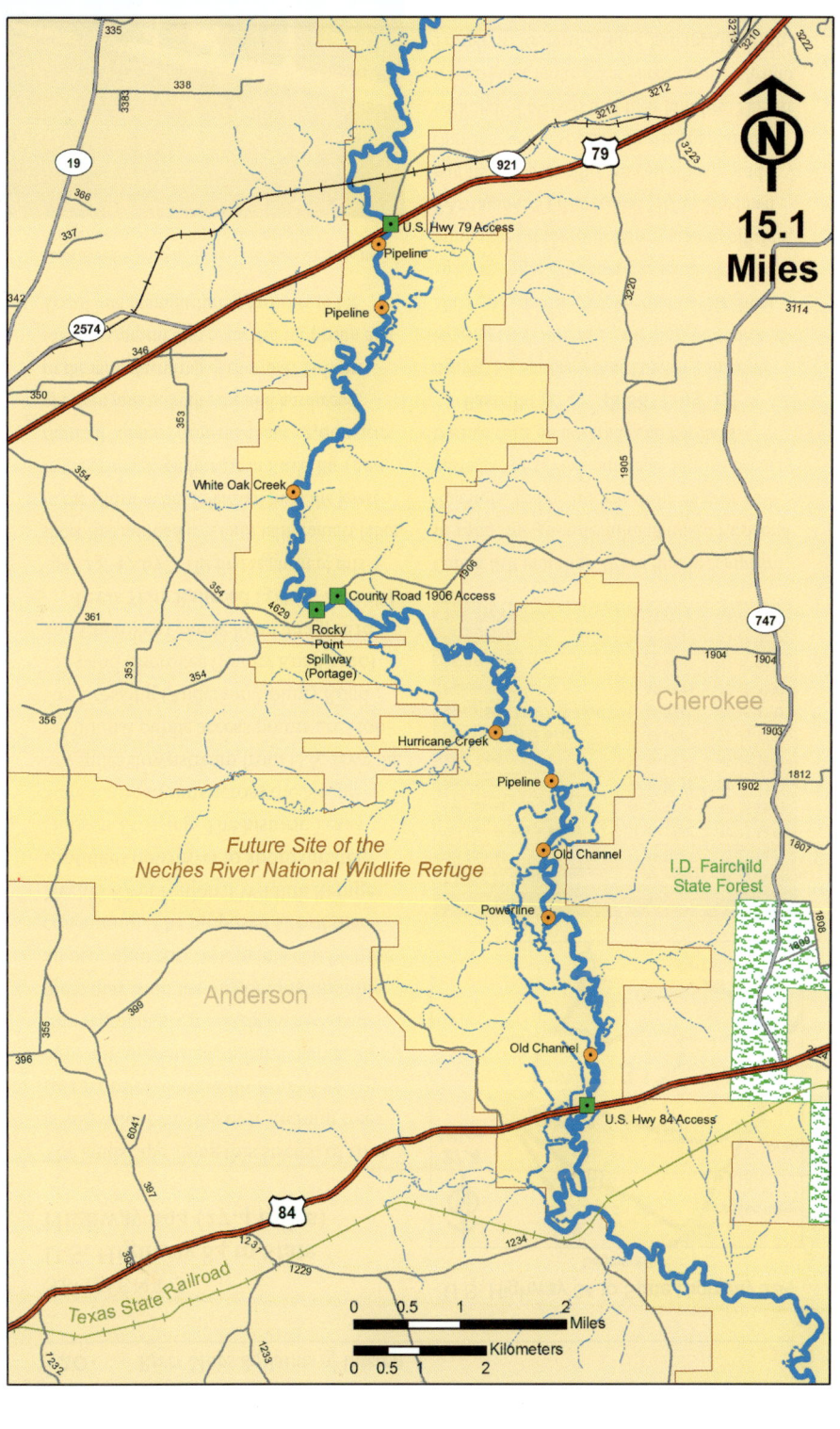

SEGMENT 3
U.S. Highway 84 to State Highway 294 (27.4 miles)

The landscape along this stretch of river is still wild, boasting over-arching trees (planer trees, commonly known as water elm) and numerous logjams. Lands along this segment are privately owned. Paddlers may want to consider making this an overnight trip, as logjams can be time consuming. The proposed site of the Neches River National Wildlife Refuge continues downriver to approximately Hobson Crossing. The historic Texas State Railroad runs across the Neches a few miles downstream from U.S. Hwy 84. Additional public access points are Hobson Crossing (Anderson County Road 1225 and Cherokee County Road 2120) and Cherokee County Road 2223/2219.

Texas State Railroad (31° 45′ 47.20″ / -95° 23′ 37.67″) *The first segment of this short-line railroad was built with convict labor in the 1880s to haul charcoal from the forest to the prison's iron furnace in Rusk. By 1900, charcoal "camps" clear-cut about 12,000 acres of virgin forests, both pine and hardwood, in the manufacturing of charcoal. This first railway segment also connected the prison and Texas' iron manufacturing facilities to the*

U.S. Highway 84 to State Highway 294
Segment 3

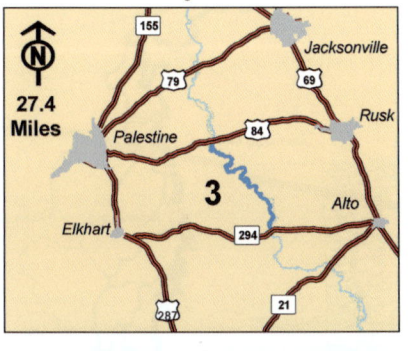

Points of Interest	Latitude	Longitude
U.S. Hwy 84 Access	31 46 36.97	-95 23 46.73
Old Channel	31 46 29.67	-95 23 56.72
Old Hwy 84 Bridge Pilings	31 46 19.90	-95 23 54.67
Old Channel	31 45 43.52	-95 24 08.45
Texas State Railroad	31 45 47.20	-95 23 37.67
Tailes Creek	31 44 03.83	-95 20 38.31
County Road 1225 Access	31 43 04.69	-95 20 32.82
County Road 2223/2219 Access	31 42 05.15	-95 18 43.28
Pipeline	31 41 49.89	-95 18 42.93
Ioni Creek	31 39 48.97	-95 17 02.23
Old Channel	31 39 40.31	-95 16 40.72
State Hwy 294 Access	31 37 48.88	-95 17 09.75

Cotton Belt mainline. When finally completed in 1909, the 30-mile railroad linked Rusk and Palestine. Today, ancient steam engines pull passenger cars filled with eager tourists between the two cities on weekends and holidays.

Fastrill (31° 37′ 48.88″ / -95° 17′ 09.75″) *A Southern Pine Lumber Company logging town thrived on the bluff just north of Hwy 294, on the Cherokee county side, from 1922 to 1941. The town was named by combining the surnames of three company officials, FArrington, STRauss, and HILL. The town had its own post office, several schools, and churches, which served a diverse population of six hundred workers and their families. The site is now home to the Texas Forest Service's Arthur Temple Sr. Research Area, a national leader in forest research science since 1951.*

Texas State Historical Railroad steam engine crossing the Neches River. Courtesy Jonathan Gerland

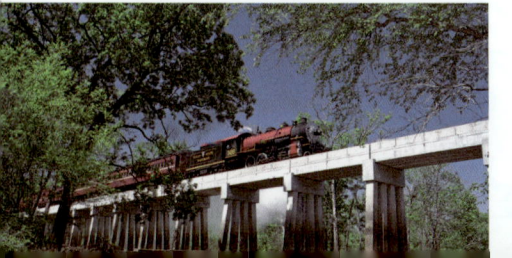

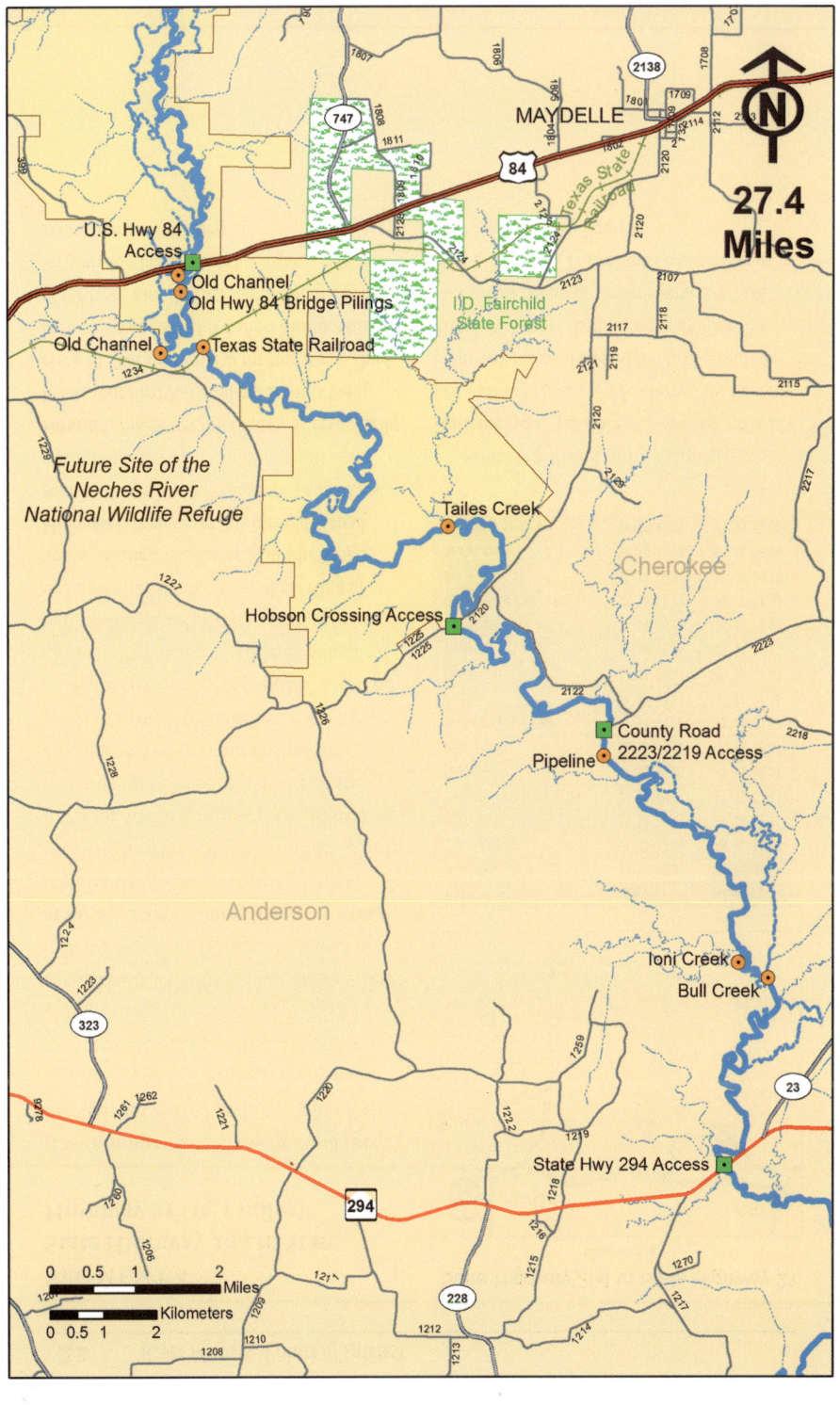

N

27.4
Miles

MAYDELLE

84

747

2138

U.S. Hwy 84 Access
Old Channel
Old Hwy 84 Bridge Pilings
Old Channel
Texas State Railroad

I.D. Fairchild State Forest

Future Site of the Neches River National Wildlife Refuge

Tailes Creek

Cherokee

Hobson Crossing Access

County Road 2223/2219 Access
Pipeline

Anderson

323

Ioni Creek
Bull Creek

23

State Hwy 294 Access

294

228

0   0.5 1   2
Miles

0 0.5 1   2
Kilometers

SEGMENT 4
State Highway 294 to State Highway 21 (15.1 miles)

It is common to hear coyotes and great horned and barred owls at night on this remote section of the Neches. There are no additional public access points along this portion of river, so plan accordingly. Camping is available a few miles west of the Neches on Hwy 21 at Mission Tejas State Historical Park, which offers water, electricity, restrooms, showers, group camping, picnic pavilion, and a dump station. Caddoan Mounds State Park is just east of the Neches on Hwy 21, but it does not provide camping facilities.

State Hwy 21 (31° 34′ 45.88″ / -95° 09′ 58.05″) *In 1691, Domingo Terán de los Ríos marked and joined together a series of Indian and game trails in the name of the King of Spain, creating El Camino Real (the King's Highway). It was over this legendary road that much of history flowed into Texas. Moses and Stephen F. Austin, Davy Crockett, Sam Houston, Jim Bowie, and many of Texas' early heroes rode their horses along what is today Hwy 21. The grave site of the first Anglo child born in Texas (1804) is a short distance east of the river, and the first settlement in Texas, mission San Francisco de los Tejas, 1690, is a stone's throw to the west.*

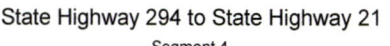
State Highway 294 to State Highway 21

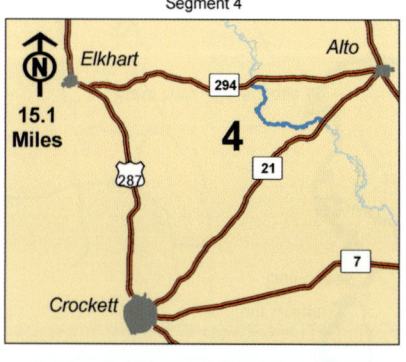
Segment 4

Points of Interest	Latitude	Longitude
State Hwy 294 Access	31 37 48.88	-95 17 08.75
Caney Creek	31 37 32.26	-95 16 55.60
Brown Branch Creek	31 37 32.43	-95 16 32.51
Proposed Fastrill Dam Site	31 37 10.06	-95 15 42.82
Powerline	31 36 46.31	-95 15 38.32
Miles Creek	31 35 43.09	-95 16 30.41
Houston County Line	31 35 34.18	-95 16 23.97
Pipeline	31 34 32.72	-95 15 21.56
Pipeline	31 34 29.31	-95 15 16.67
Powerline	31 34 30.26	-95 14 41.72
San Pedro Creek	31 34 26.94	-95 13 53.91
Old Railroad Bridge Pilings	31 34 46.34	-95 10 56.52
Box Creek	31 34 50.32	-95 10 50.16
Bowles Creek	31 34 48.17	-95 10 03.46
State Hwy 21 Access	31 34 45.88	-95 09 58.05

Southern Pine Lumber Company Railroad (31° 34′ 46.34″ / -95° 10′ 56.52″) *Just north of Hwy 21, a Southern Pine Lumber Company railroad crossed the Neches and connected with more than 60 miles of mainline that linked Diboll to timber reserves on both sides of the river. The railroad was known as "The Neches Valley Route."*

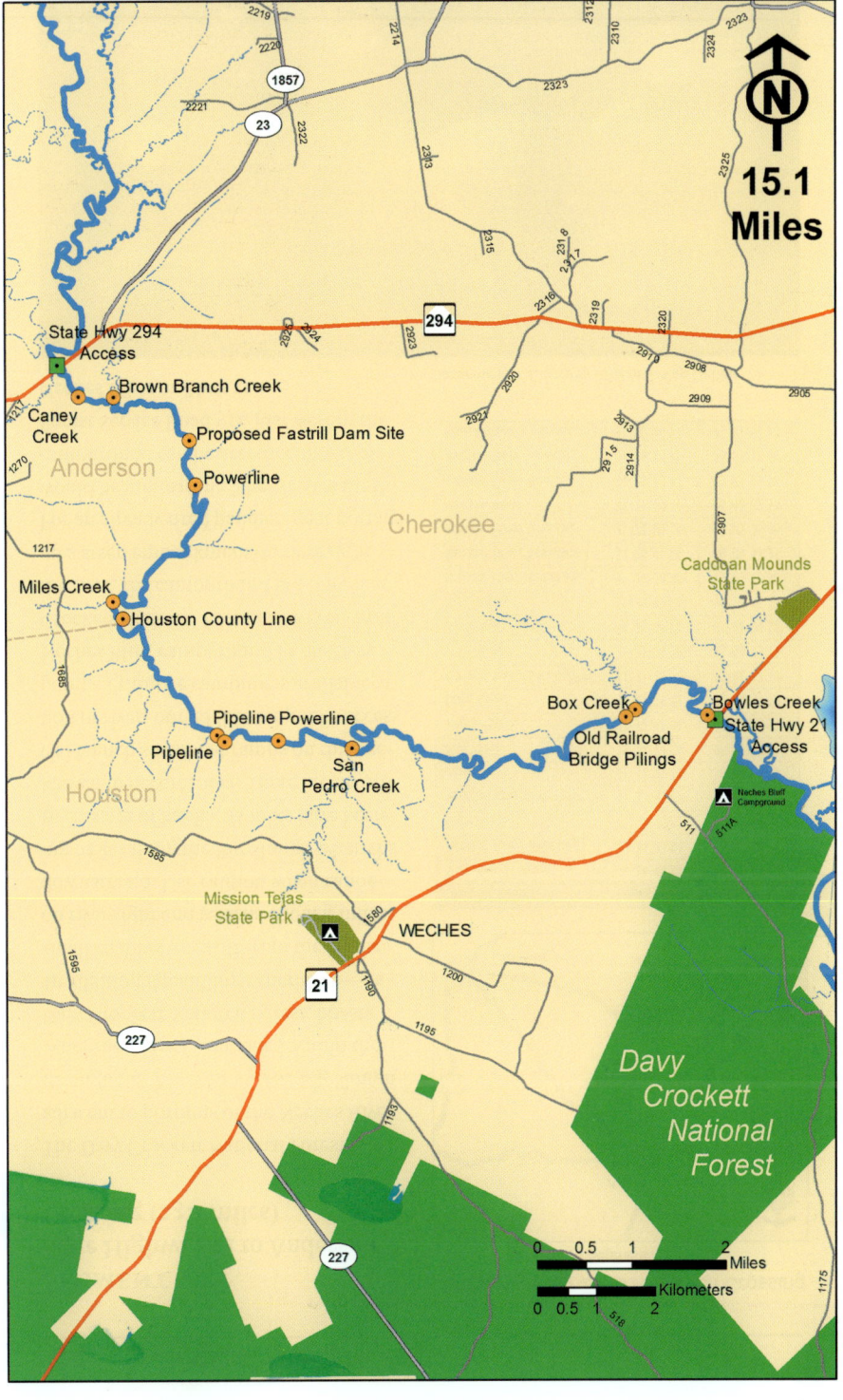

SEGMENT 5
State Highway 21 to Anderson Crossing (23.1 miles)

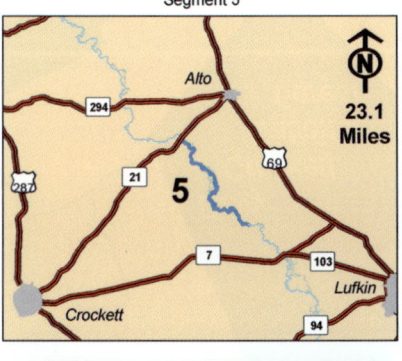

State Highway 21 to Anderson Crossing
Segment 5

23.1 Miles

The Davy Crockett National Forest is adjacent to portions of the Neches River throughout this section. The Big Slough Wilderness Area, contained within the Davy Crockett National Forest, boasts an additional paddling experience—its small channel goes off from the Neches on river right and returns about 4 miles downstream. Use caution when canoeing or kayaking the Big Slough when the river is out of banks, as folks have been known to become lost. There are no official launch sites/boat ramps on this section of river, nor additional public access points. Primitive camping is available at Neches Bluff campground in the Davy Crockett National Forest; however, from the river, the campground is a long hike up a steep embankment on river right. For an already tired paddler, gear portage would be unpleasant. The Neches Bluff campground can be accessed by car on Forest Service Road 511, just west of the Neches River on Hwy 21.

Points of Interest	Latitude	Longitude
State Hwy 21 Access	31 34 45.88	-95 09 58.05
National Forest Boundary	31 34 31.64	-95 09 44.18
Cedar Creek	31 33 49.83	-95 08 48.41
National Forest Boundary	31 33 45.97	-95 08 39.06
Pipeline	31 32 27.71	-95 08 40.04
Bluff Creek	31 32 24.66	-95 08 37.91
Sulphur Creek	31 32 17.94	-95 07 36.97
Camp Creek	31 31 28.42	-95 06 37.18
National Forest Boundary	31 31 24.24	-95 06 17.79
Old Channel	31 31 24.28	-95 06 11.73
Big Slough	31 30 44.83	-95 06 00.50
Big Slough	31 29 07.11	-95 06 35.72
National Forest Boundary	31 28 28.68	-95 05 55.56
National Forest Boundary	31 28 06.76	-95 05 53.24
National Forest Boundary	31 28 02.19	-95 05 29.40
Larrison Creek	31 26 54.74	-95 02 52.37
National Forest Boundary	31 26 42.17	-95 02 42.78
National Forest Boundary	31 26 44.95	-95 05 15.69
Anderson Crossing Access	31 26 40.97	-95 02 03.18

Canoe campers. Courtesy Adrian F. Van Dellen

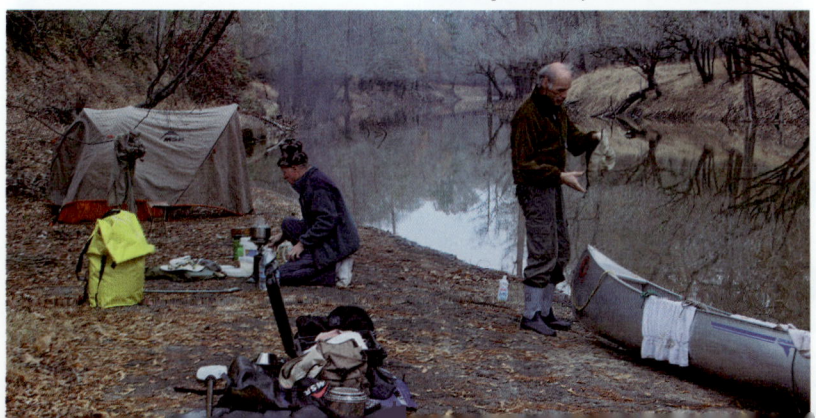

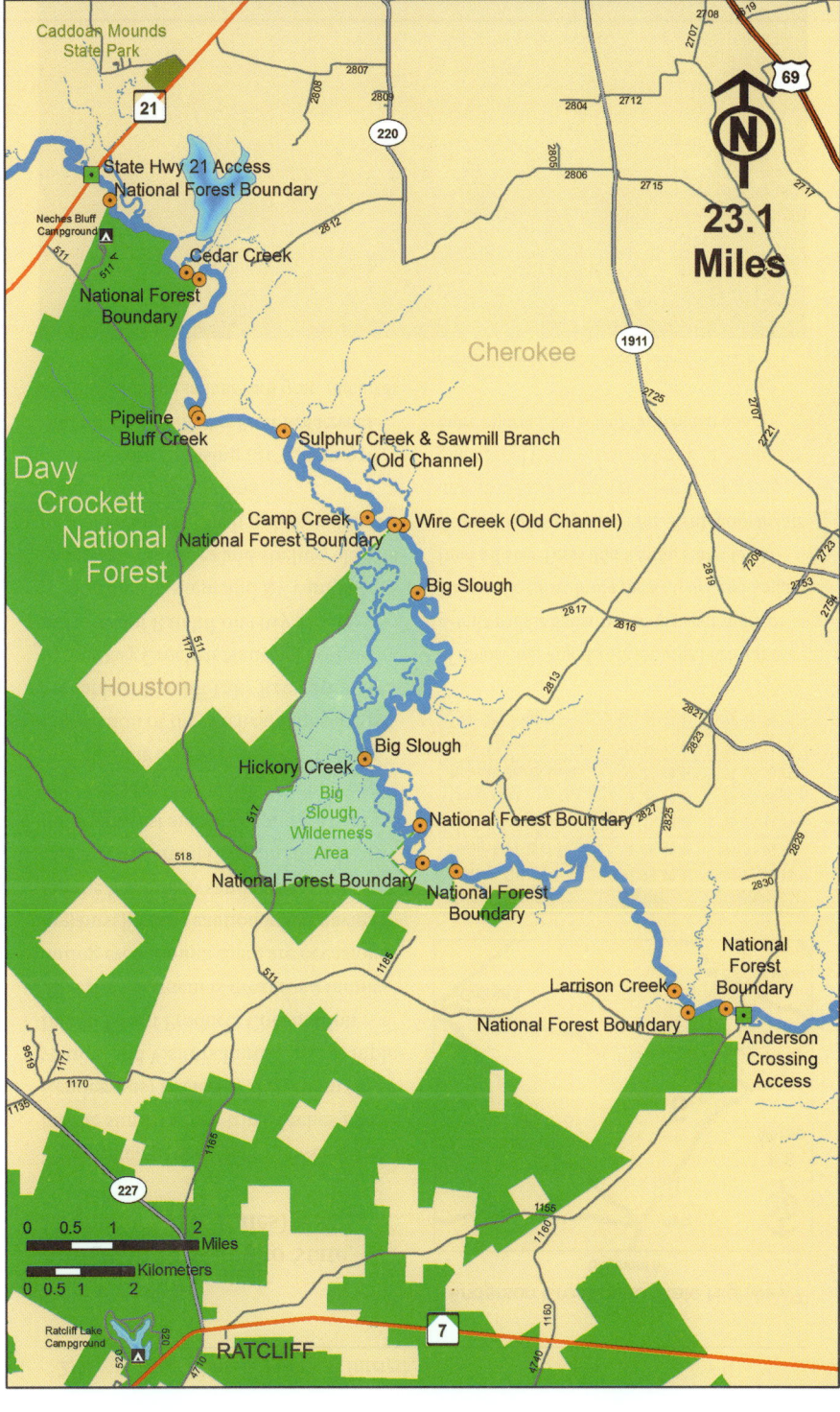

SEGMENT 6
Anderson Crossing to State Highway 7 (8.8 miles)

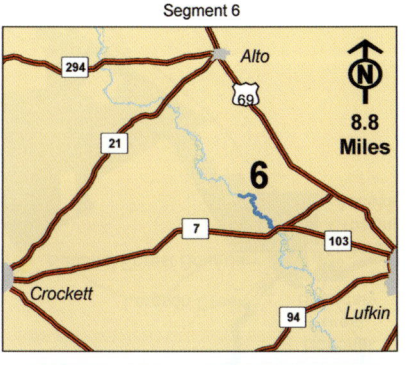

Anderson Crossing to State Highway 7
Segment 6

This stretch of the Neches is well known for the annual canoe trip (Neches River Rendezvous) hosted by the Lufkin Convention and Visitors Bureau and Temple-Inland Forest Products Corporation. This river segment is great for a family outing even though there are occasional logjams; just use caution when navigating the hazards. A sandy clay access site exists at Anderson Crossing bridge on the southwest side of the river. There is a public boat ramp at Hwy 7, but no additional public access points exist along this section of the Neches. Camping is available at Ratcliff Lake Recreation Area in the Davy Crockett National Forest 1 mile west of Ratcliff on Hwy 7. The recreation area has tent and RV campsites, group camping area and shelter, electricity, cold showers, flush toilets, a dump station, and concessions.

Anderson Crossing (31° 26' 40.97" / -95° 02' 03.18") *Neches River Rendezvous— The first Saturday in June each year, hundreds*

Points of Interest	Latitude	Longitude
Anderson Crossing Access	31 26 40.97	-95 02 03.18
County Road 2829 Access	31 26 40.46	-95 02 02.98
Angelina County Line	31 25 32.66	-95 00 11.94
Conner Creek	31 25 17.77	-95 00 20.44
McCombe Branch Creek	31 25 08.46	-95 00 10.05
Devils Bayou	31 24 31.45	-94 58 35.36
State Hwy 7 Access	31 23 47.67	-94 57 55.54

of canoes and kayaks are launched here for the 9-mile paddle downriver to the boat ramp at Hwy 7. Paddlers bring their canoes and kayaks from all over Texas and nearby states to participate in this popular event. Rental canoes are also available. Contact the Angelina County Chamber of Commerce at (936) 634-6644 for more information about the Rendezvous.

Neches River rendezvous. Courtesy Mark W. Bush

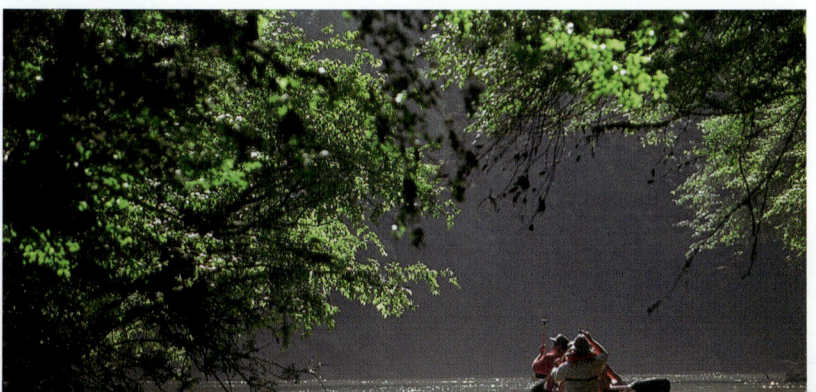

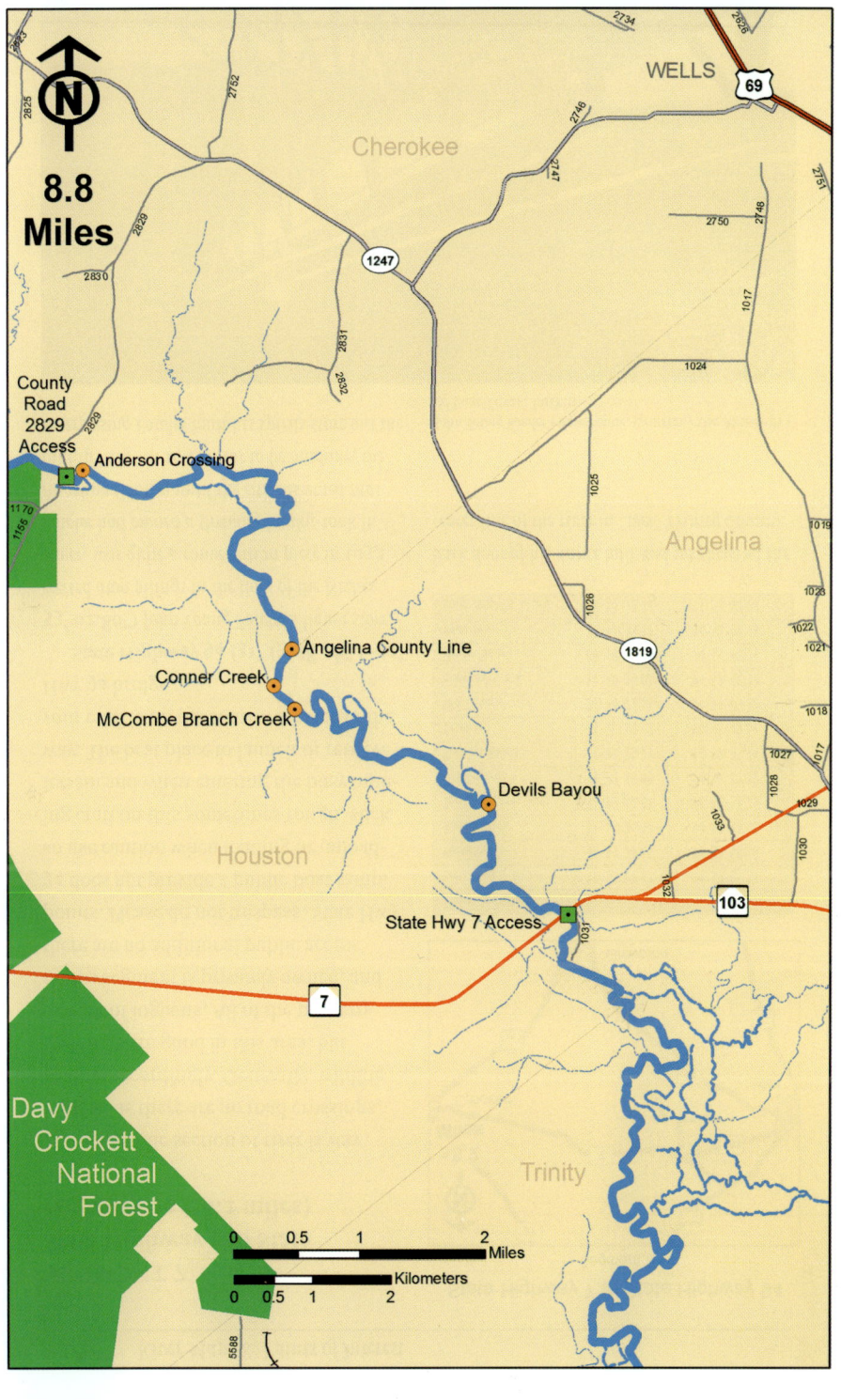

SEGMENT 7
State Highway 7 to State Highway 94 (18.2 miles)

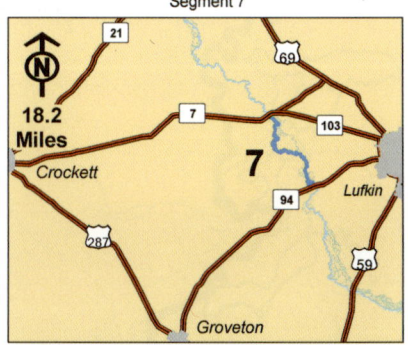

State Highway 7 to State Highway 94
Segment 7

This 18.2-mile section of river is very remote, as there are no road crossings, so plan accordingly. Generally, wildlife sightings are good in this area, but beware of logjams. All of the property in this segment is privately owned, and there are no additional public access points. Please do not trespass. State Hwy 94 does not provide a public boat ramp, so use caution when loading or unloading craft on this sometimes rough, slick terrain and when entering the busy highway. The best place to launch or retrieve your craft is on the southwest side of the Hwy 94 bridge.

State Highway 94 (31° 17′ 18.76″ / -94° 53′ 02.80″) *John Young Fowler's liquor store, erected atop pilings in the flow of the Neches River, was quite a conversation piece in 1937. Fowler had owned a thriving honky-tonk in Angelina County until the citizens voted that county "dry" in 1936. Not to be outdone, the enterprising Fowler built his spirits store out the back door of his dance hall and just beyond the center line of the river in "wet" Trinity County.*

Points of Interest	Latitude	Longitude
State Hwy 7 Access	31 23 47.67	-94 57 55.54
Powerline	31 23 28.15	-94 57 58.16
Bodan Creek	31 23 20.59	-94 57 33.65
Trinity County Line	31 23 12.87	-94 57 28.79
Old Channel	31 22 46.09	-94 57 11.22
Cochino Bayou	31 20 25.21	-94 57 03.24
Pipeline	31 20 16.12	-94 55 20.07
Old Channel	31 20 22.13	-94 55 03.36
Crawford Creek	31 19 25.03	-94 53 48.32
Boggy Slough	31 18 02.95	-94 53 48.21
Powerline	31 17 23.32	-94 53 02.46
State Hwy 94 Access	31 17 18.76	-94 53 02.80

John Young Fowler's liquor store. Courtesy the Museum of East Texas, Lufkin

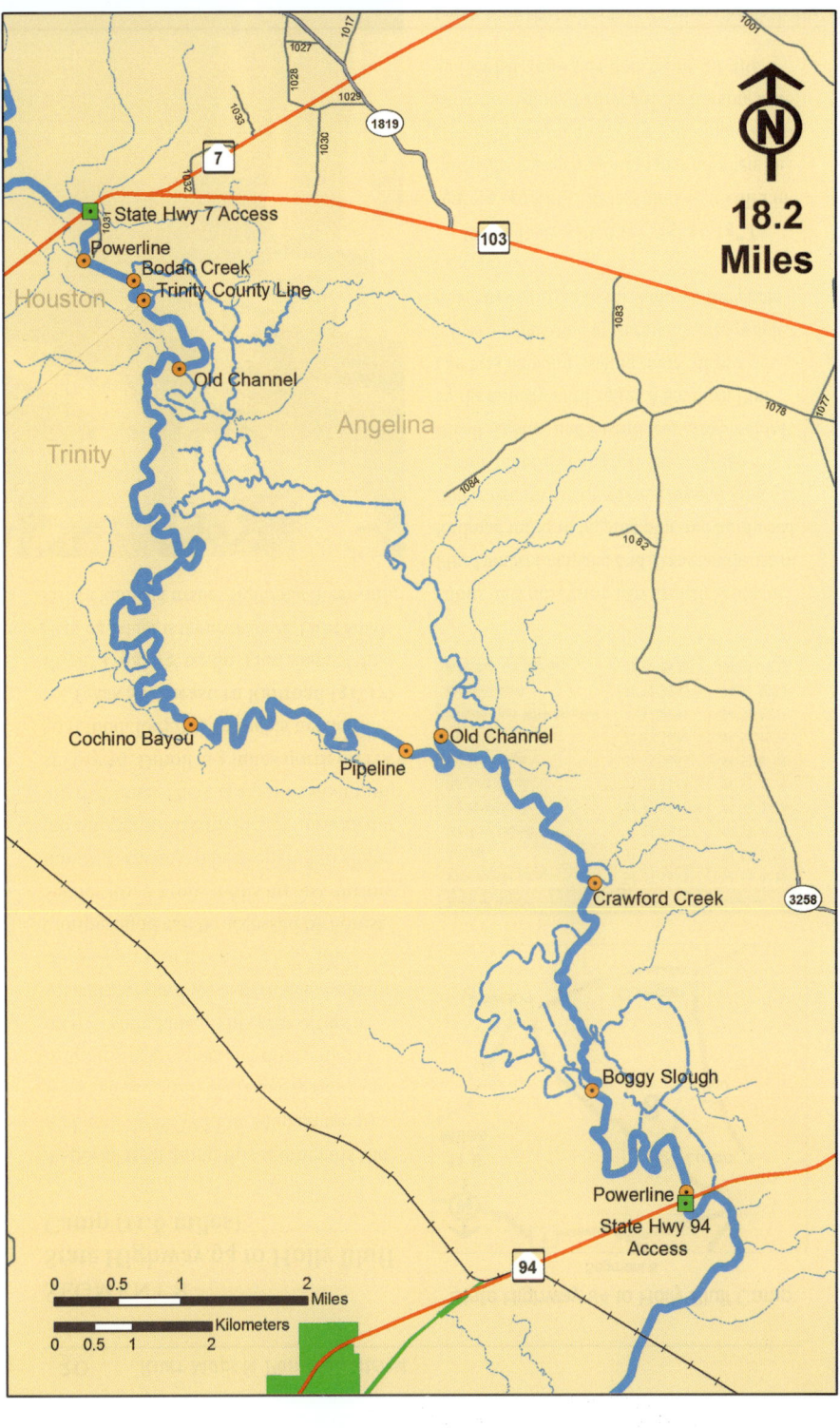

N

18.2
Miles

1017

1027

1028

1029

1033

1030

7

1819

1031

103

1032

Houston

State Hwy 7 Access
Powerline
Bodan Creek
Trinity County Line

Old Channel

1083

Angelina

1078

1077

Trinity

1084

1082

Cochino Bayou

Old Channel

Pipeline

Crawford Creek

3258

Boggy Slough

Powerline
State Hwy 94
Access

94

0   0.5   1   2
                                    Miles
                        Kilometers
0   0.5   1   2

SEGMENT 8
State Highway 94 to Holly Bluff Camp (11.6 miles)

A special treat along this segment is the Alabama Creek Wildlife Management Area and the Holly Bluff Camp on river right. If you are quiet, wildlife can be seen in abundance. This area offers a spectacular outdoor experience. Logjams can still present hazards. Holly Bluff Campground can be accessed by Forest Service Road 510A. It fills up fast during hunting season, so be prepared to share the campground with lots of hunters during that time. There is a public boat ramp at Hwy 59. Diboll is 3 miles north on Hwy 59 if food, lodging, or fuel is needed.

Texas Southeastern Railroad (31° 17′ 18.76″ / -94° 53′ 02.80″) *Just south of the Hwy 94 bridge is the abandoned Texas Southeastern Railroad trestle. Beginning in the early*

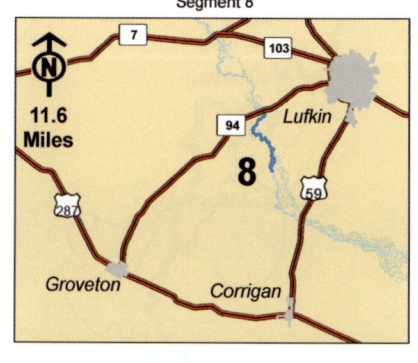

State Highway 94 to Holly Bluff Camp
Segment 8

Points of Interest	Latitude	Longitude
State Hwy 94 Access	31 17 18.76	-94 53 02.80
Railroad	31 16 25.43	-94 53 21.93
Old River Channel	31 15 55.13	-94 53 49.54
Big Bend Slough	31 15 45.12	-94 53 45.76
Old River Channel	31 15 12.64	-94 53 28.79
Big Bend Slough	31 15 20.36	-94 53 07.39
Cross Slough	31 13 56.02	-94 51 45.51
Alabama Creek WMA Boundary	31 13 09.17	-94 51 44.93
Hackberry Creek	31 12 33.57	-94 51 59.56
Holly Bluff Camp Access	31 12 05.20	-94 51 46.29

1900s, this rail system was used by Southern Pine Lumber Company and Texas Southeastern Railroad trains in tapping pine and hardwood timber resources in Cherokee, Houston, and Trinity counties. Passenger service between Diboll, Lufkin, and Fastrill was also provided by the railway until 1942. A Southern Pine Lumber Company logging camp, Alcedo, sustained a population of several hundred people from 1915 to 1924 at this location (see photo page 32).

Black Bear Killed (31° 15′ 55.13″ / -94° 53′ 49.54″) *The last black bear killed in the Neches bottomlands was shot near here on Christmas Eve, 1962, by Joe Harber and Henry Lawrence. Young Joe dropped the running animal with a single shot from a .30-30 rifle.*

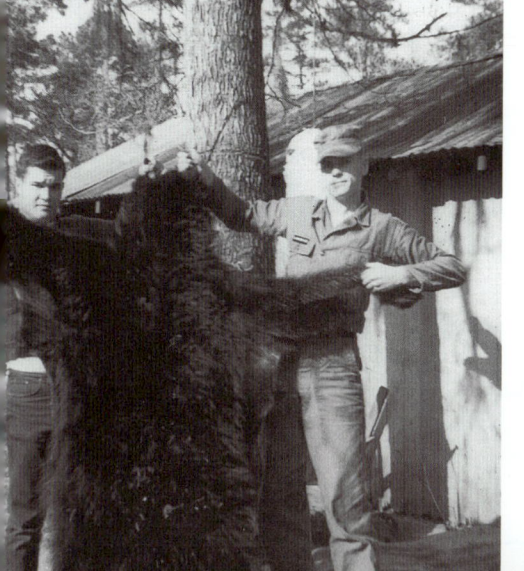

Last black bear killed in East Texas. Courtesy Charlie Harber

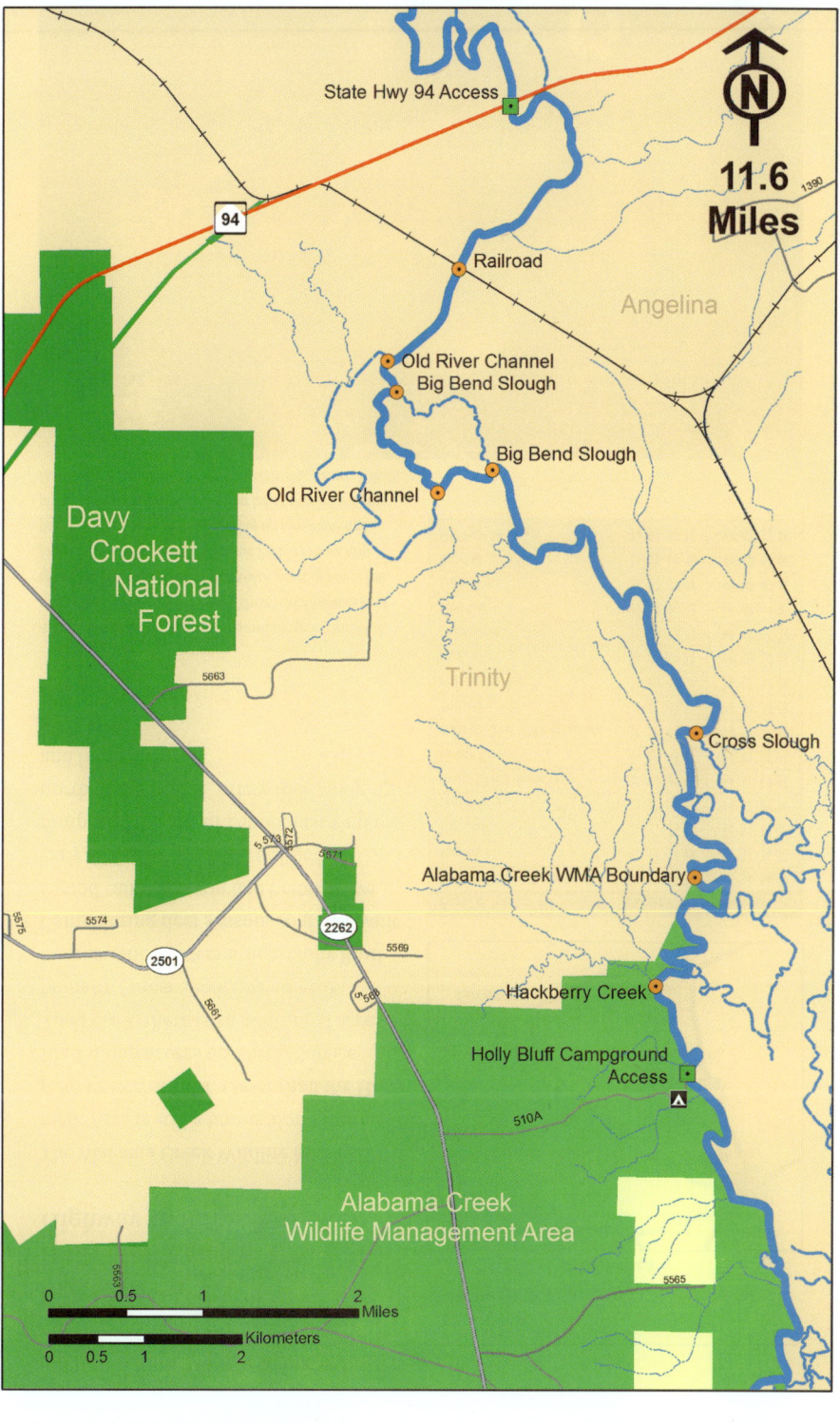

State Hwy 94 Access

11.6 Miles

1390

94

Railroad

Angelina

Old River Channel
Big Bend Slough

Big Bend Slough

Old River Channel

Davy
Crockett
National
Forest

5663

Trinity

Cross Slough

5 573
5572
5571

Alabama Creek WMA Boundary

5575

5574

2262

5569

2501

5661

5568

Hackberry Creek

Holly Bluff Campground
Access

510A

Alabama Creek
Wildlife Management Area

5663

5565

| 0 | 0.5 | 1 | | 2 |
Miles

| 0 | 0.5 | 1 | | 2 |
Kilometers

SEGMENT 9
Holly Bluff Camp to U.S. Highway 59 (9.1 miles)

The Alabama Creek Wildlife Management Area is still a bonus along this 9.1-mile stretch of river. Other than the Holly Bluff Camp access on Forest Service Road 510A, there is an additional access point on Forest Service Road 510B. Enthusiastic hunters utilize Holly Bluff Camp during deer season, so if you want a good campsite, you had better arrive early. U.S. Hwy 59 has a public boat ramp, and the city of Diboll, just 3 miles north of the bridge, offers food, lodging, and fuel.

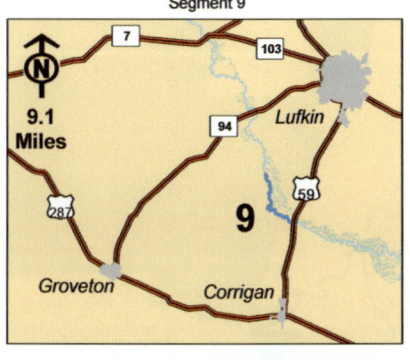

Holly Bluff Camp to U.S. Highway 59
Segment 9

Points of Interest	Latitude	Longitude
Holly Bluff Camp Access	31 12 05.20	-94 51 46.29
Pipeline	31 11 56.39	-94 51 36.36
Slay Creek	31 11 25.18	-94 51 27.38
Powerline	31 11 01.34	-94 51 09.54
Forest Service Road 510B	31 10 21.59	-94 51 19.14
Carlton Creek	31 09 49.11	-94 51 08.90
Alabama Creek WMA Boundary	31 09 16.06	-94 51 15.69
Alabama Creek	31 08 47.51	-94 50 34.48
Polk County Line	31 08 47.51	-94 50 34.48
Bluffs	31 08 30.27	-94 50 12.87
Cedar Creek	31 08 14.74	-94 49 15.74
Powerline	31 08 00.33	-94 48 41.92
Clarks Ferry	31 07 59.97	-94 48 47.11
Railroad	31 08 00.48	-94 48 43.40
U.S. Hwy 59 Access	31 07 58.15	-94 48 37.48

A Texas Southeastern Railroad log train approaches Diboll with a load of logs, about 1907. These steam engines were often fired with "pine knots" (resin-rich pieces of pine). Jim Jones and his father, of the Shawnee Prairie community, paid for a forty-acre farm by picking up pine knots and piling them alongside the rails that led to the Carter-Kelly Lumber Company mill at Manning. (History Center, Diboll)

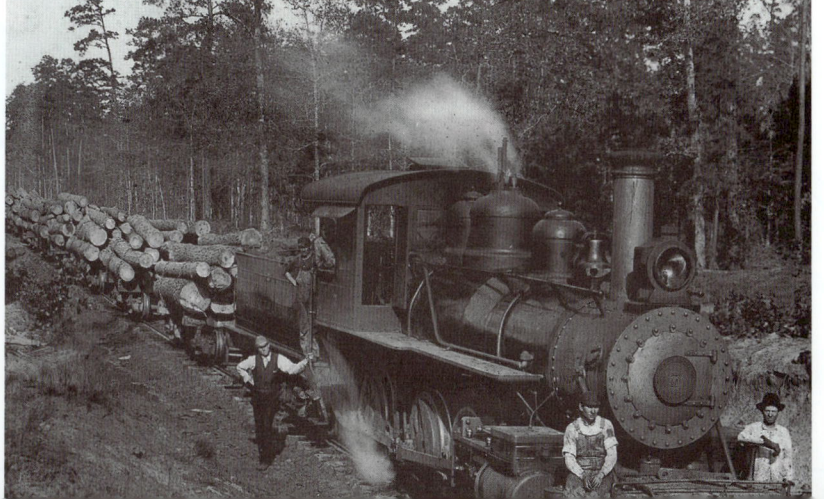

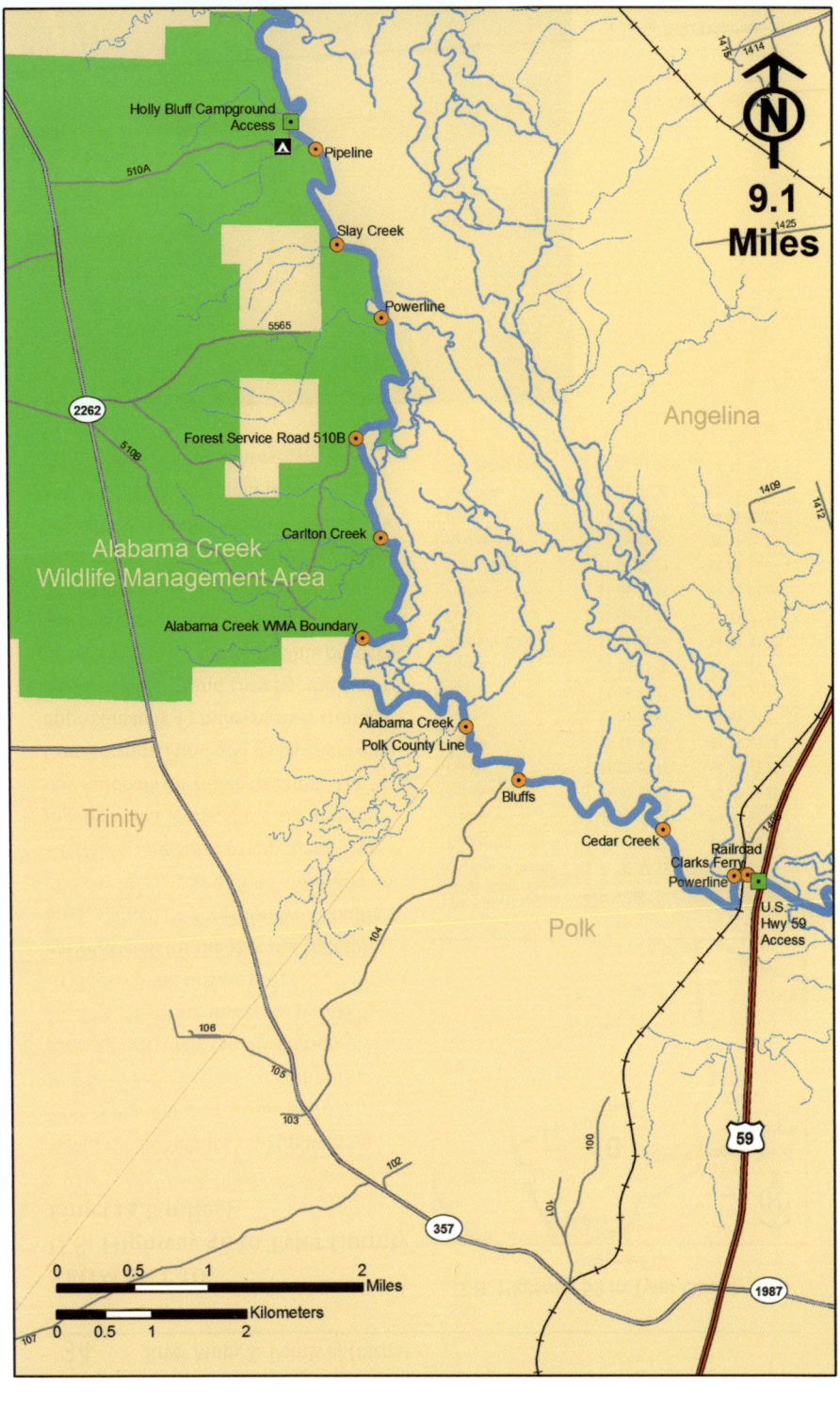

SEGMENT 10
U.S. Highway 59 to Tyler County Line (24.8 miles)

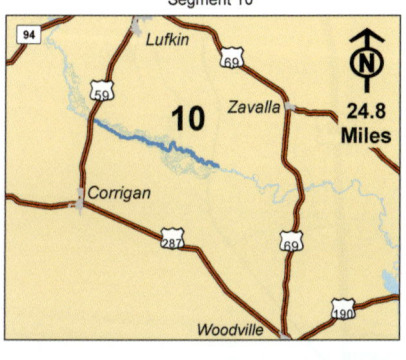

When you launch at U.S. Highway 59, be prepared for a 43.3 mile journey; this segment and the next combine to form an extremely remote section of the Neches. There are no access points by road, except on private property, in this entire stretch (or the 18.5 mi. that follow), so prepare accordingly. A couple of days' provisions of food and water are advisable. Logjams are abundant. During low water levels, there are numerous sandbars on which to camp. The Conservation Fund and its partners own approximately 33 miles of river frontage between Hwy 59 and Hwy 69, the majority of which is on river left. This property has been designated as the Pineywoods Mitigation Bank, and when all credits from the bank have been sold, the property will be transferred to the Texas Parks and Wildlife Department for use as a Wildlife Management Area.

Points of Interest	Latitude	Longitude
U.S. Hwy 59 Access	31 07 58.15	-94 48 37.48
Pipeline	31 07 51.21	-94 48 20.43
Pipeline	31 07 51.60	-94 47 57.81
Dollarhide Cutoff	31 07 42.98	-94 47 36.25
Pipeline	31 07 05.17	-94 46 32.07
Pipeline	31 06 48.05	-94 45 47.88
Pipeline	31 06 35.63	-94 45 24.27
Pipeline	31 06 34.22	-94 45 22.35
Pipeline	31 06 33.78	-94 45 19.37
Dollarhide Cutoff	31 06 21.59	-94 43 42.42
Pipeline	31 06 29.17	-94 43 27.01
Pipeline	31 06 23.44	-94 43 03.86
Lamar Slough	31 05 52.59	-94 41 26.89
Pipeline	31 05 41.02	-94 39 15.43
Pipeline	31 05 48.88	-94 39 07.88
Biloxi Creek	31 04 34.20	-94 36 14.87
Tyler County Line	31 03 31.54	-94 33 43.53

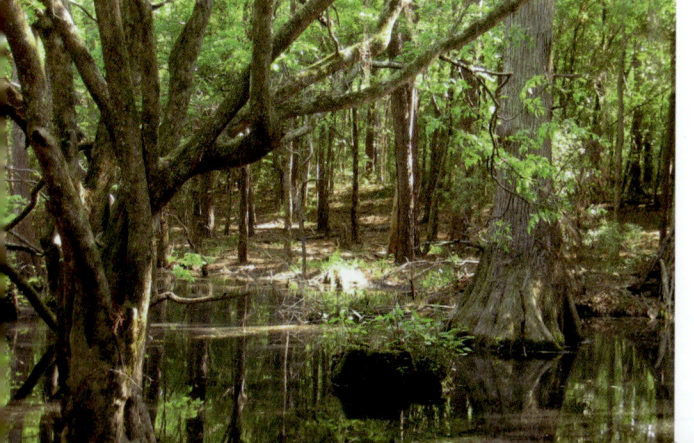

Neches River bottomlands provide food and shelter for numerous species of birds and wildlife.
Courtesy Gina Donovan

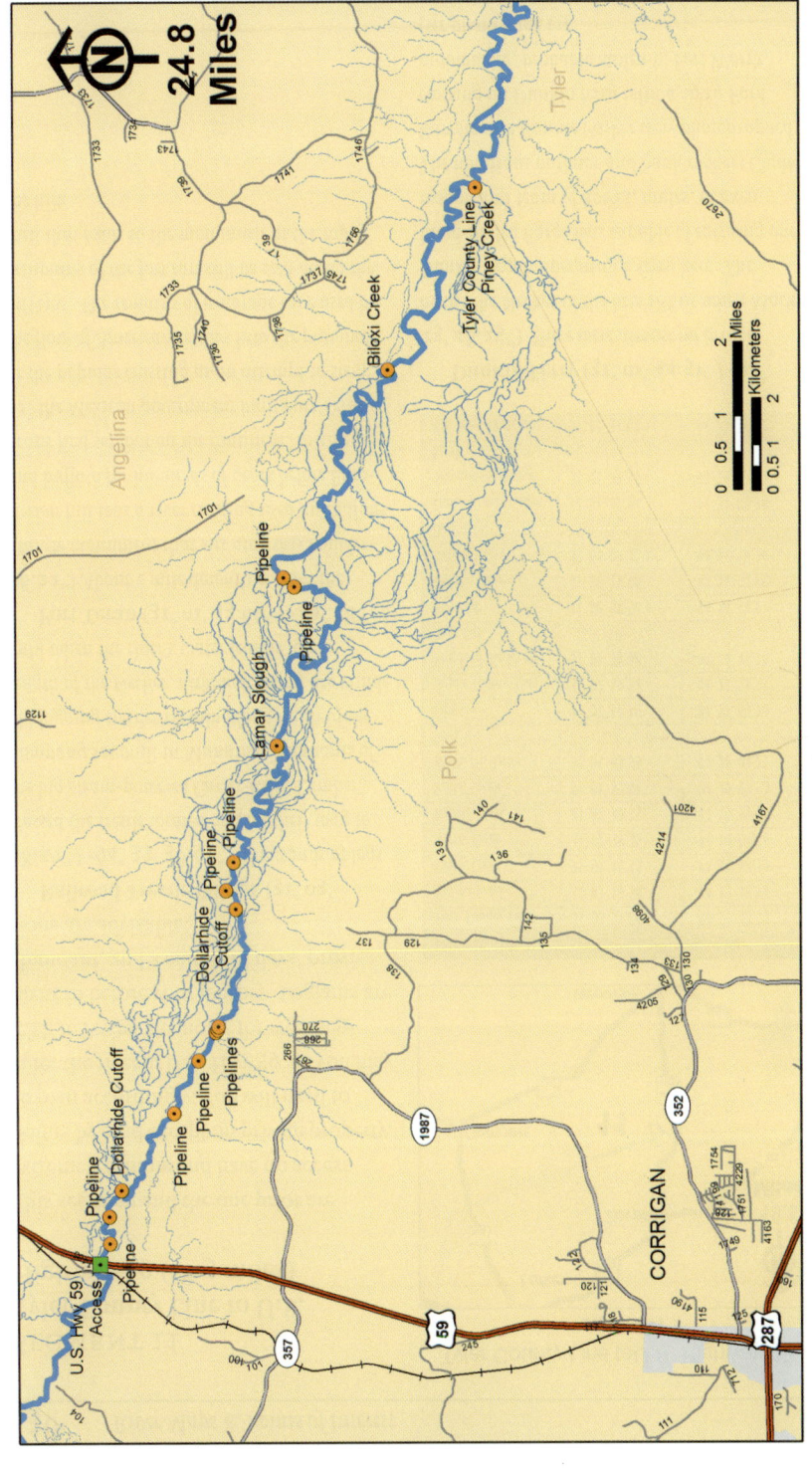

SEGMENT 11
Tyler County Line to U.S. Highway 69 (18.5 miles)

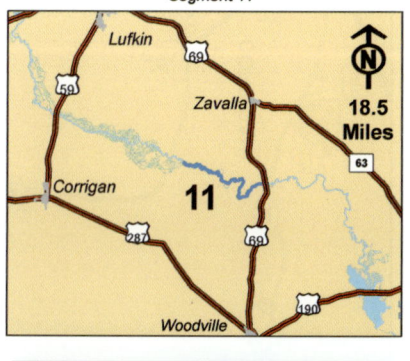

Tyler County Line to U.S. Highway 69
Segment 11

This segment and the one prior are extremely remote and have no access points by road except on private property, so plan accordingly. You will need to enter the river at U.S. Hwy 59, so you may wish to review the notes for river segment 10 before your journey. Logjams are abundant, and a couple of days' provisions are advisable.

Railroad Trestle Pilings (31° 03′ 28.56″ / -94° 33′ 36.97″) *Thousands of logs crossed the trestle located here on their way to the big steam-powered Carter-Kelly Lumber Company sawmill in Manning. Remnants of old logging trestles remain along much of the length of the Neches, although many are visible only when the river's water level is low.*

Fort Teran (31° 01′ 43.66″ / -94° 28′ 06.22″) *About 1 mile downriver from the granite monument that sits atop this boulder-strewn hill was a river crossing used by Indians and buffalo for hundreds of years before Europeans first set foot on the continent. In 1830–31, the Mexican government built Fort Teran at the popular crossing as an attempt to stem the flow of American settlers into the province of Tejas. The remains of a storage cave used by occupants of the fort can still be seen along the trail that leads to the monument at the top of the hill.*

Points of Interest	Latitude	Longitude
Tyler County Line	31 03 31.54	-94 33 43.53
Old Railroad	31 03 28.56	-94 33 36.97
Caney Creek	31 03 14.58	-94 32 57.34
Burris Creek	31 02 39.35	-94 30 04.06
Russell Creek	31 02 30.75	-94 29 59.07
Saline Creek	31 01 53.81	-94 29 09.53
Bluffs (Ft. Teran Site)	31 01 43.66	-94 28 06.22
Pipeline	31 01 54.18	-94 27 57.20
Jasper County Line	31 01 59.86	-94 27 29.23
Proposed Rockland Dam Site	31 01 23.08	-94 26 45.76
Big Bluff	31 00 09.56	-94 26 58.91
Billiams Creek	31 00 24.90	-94 26 02.98
National Forest Boundary	31 01 39.21	-94 24 14.29
Crane Branch Creek	31 01 37.37	-94 24 08.98
Greenwood Branch Creek	31 01 32.55	-94 24 07.12
U.S. Hwy 69 Access	31 01 31.02	-94 23 59.52

Dunkin Ferry (31° 01′ 44.51″ / -94° 23′ 48.25″) *There were almost no bridges across the Neches in the late 1880s when Mack Dunkin began operating a ferry here. The hand-pulled barge was capable of carrying one wagon and team of horses, mules, or oxen. The notorious outlaws and bank robbers Clyde Barrow and Bonnie Parker were photographed crossing on Dunkin ferry with a 1929 Ford automobile, probably stolen in Fort Worth. (see photo page 11)*

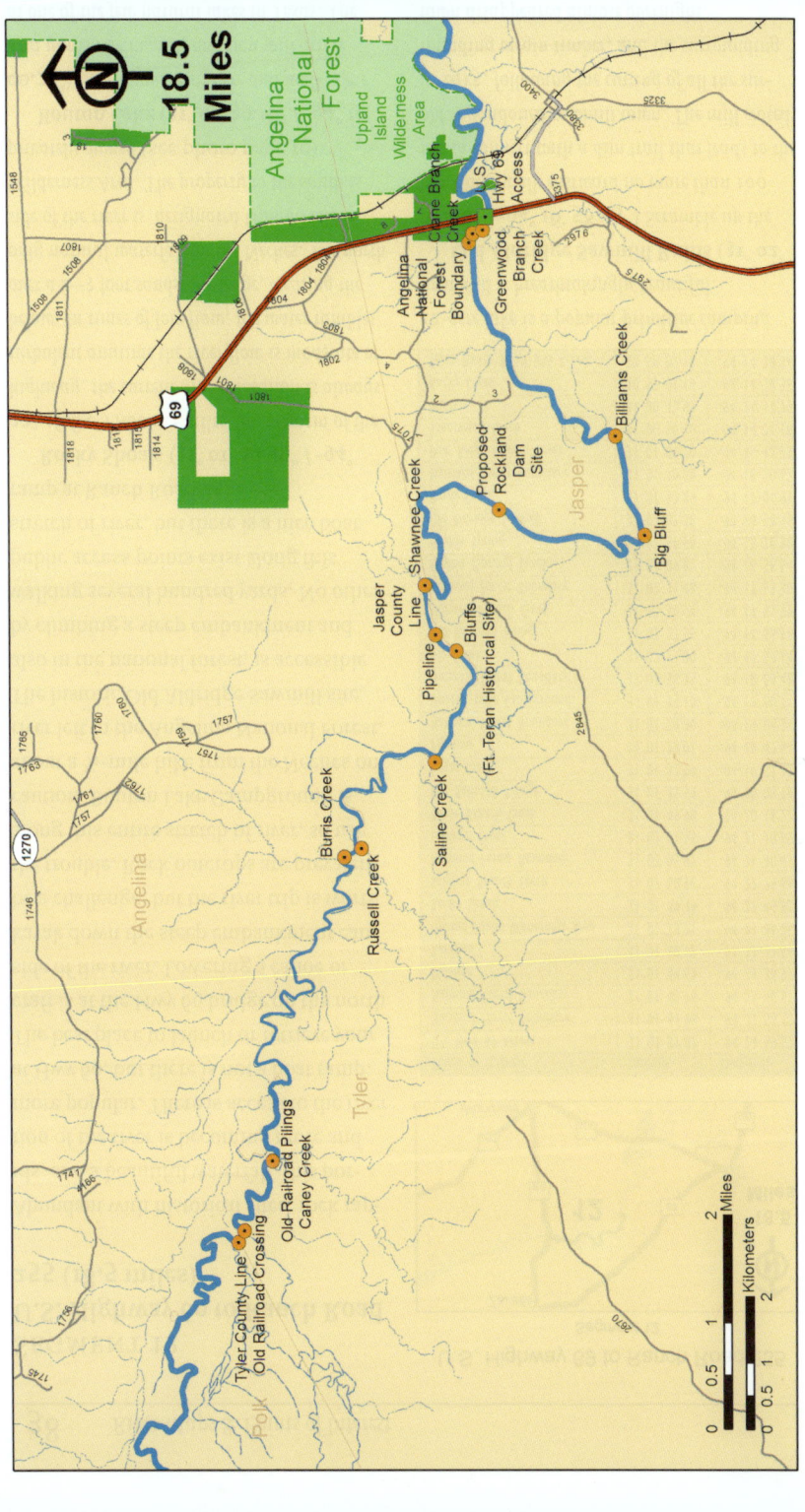

SEGMENT 12
U.S. Highway 69 to Ranch Road 255 (18.5 miles)

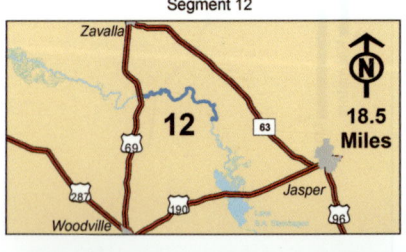

U.S. Highway 69 to Ranch Road 255
Segment 12

Abundant with historical sites, rock rapids, and a beautiful waterfall, this portion of the river is becoming more and more popular. There is access to the river at Hwy 69, but there is not a boat ramp. The best place to launch or retrieve your craft is at the Hwy 69 bridge on the north side of the river. Lowering a canoe or kayak down the steep embankment can be a challenge, but the river trip is worth the trouble. Rock outcrops are prevalent along this entire stretch of river, so use caution. Bouton Lake Campground is about a ½-mile hike from the Neches on river left in the Angelina National Forest. The historic Old Aldridge Sawmill site, also in the national forest, is accessible by climbing a steep embankment and walking several hundred yards. No other public access points exist along this stretch of river, but there is a nice boat ramp at Ranch Road 255.

Rocky Shoals (31° 01′ 49.49″ / -94° 22′ 46.83″) *About 2 miles downstream of the highway, the current at this location is always turbulent anytime the river flow is moderate or below. In times of low flow, the water tumbles over a 2–3 foot sandstone ledge, creating the only natural waterfall on the Neches. The north side of the river is designated Upland Island Wilderness Area. The property to the south is privately owned (see photo page 10).*

Bouton Lake (31° 01′ 39.18″ / -94° 19′ 00.71″) *Beach your craft here and walk some 400 yards northward, and you will arrive at one of the few natural lakes in Texas. The*

Points of Interest	Latitude	Longitude
U.S. Hwy 69 Access	31 01 31.02	-94 23 59.52
National Forest Boundary	31 01 41.98	-94 23 50.13
National Forest Boundary	31 01 45.76	-94 23 46.61
Pipeline	31 01 44.69	-94 23 38.97
Railroad	31 01 38.18	-94 23 27.56
Upland Island Wilderness Area	31 01 38.21	-94 23 26.25
Rocky Shoals	31 01 49.49	-94 22 46.83
Sulphur Branch Creek	31 01 50.96	-94 22 29.66
National Forest Boundary	31 02 05.07	-94 21 30.43
Graham Creek	31 02 11.56	-94 21 18.68
Beale Branch Creek	31 01 49.98	-94 20 15.22
Old Railroad Pilings	31 01 45.54	-94 20 08.76
Sugar Creek	31 01 32.99	-94 19 57.35
Pipeline	31 01 28.01	-94 19 47.94
National Forest Boundary	31 01 24.20	-94 19 04.31
Bouton Lake Campground	31 01 39.18	-94 19 00.71
National Forest Boundary	31 01 56.35	-94 18 08.48
Big Creek	31 02 13.00	-94 17 54.14
Old Railroad Pilings	31 02 27.56	-94 17 56.96
Millseed Branch Creek	31 02 30.94	-94 17 55.99
National Forest Boundary	31 02 31.48	-94 17 53.30
Boykin (Spring) Creek	31 02 09.82	-94 16 56.54
Boykin Spring	31 03 29.58	-94 16 28.38
Old Aldridge Sawmill	31 02 07.98	-94 16 58.45
Rawls Creek	31 01 22.85	-94 17 01.91
National Forest Boundary	31 01 10.88	-94 16 10.67
Pole Ridge Branch Creek	31 01 09.20	-94 15 12.73
Shearwood Creek	31 00 50.10	-94 14 21.18
Bluffs	31 00 42.09	-94 14 17.79
Rocky Creek	30 59 01.18	-94 14 38.36
Texas Ranch Road 255 Access	30 59 01.18	-94 14 38.36

12-acre lake is a popular primitive camping area and is breathtakingly beautiful.

Old Aldridge Sawmill Ruins (31° 02′ 07.98″ / -94° 16′ 58.45″) *Scramble up the bank and walk eastward no more than 100 feet to connect with a dim trail that leads to the old, abandoned sawmill town. The mill closed in 1915, following the cutting of all the surrounding virgin timber, and the surrounding town disappeared almost overnight.*

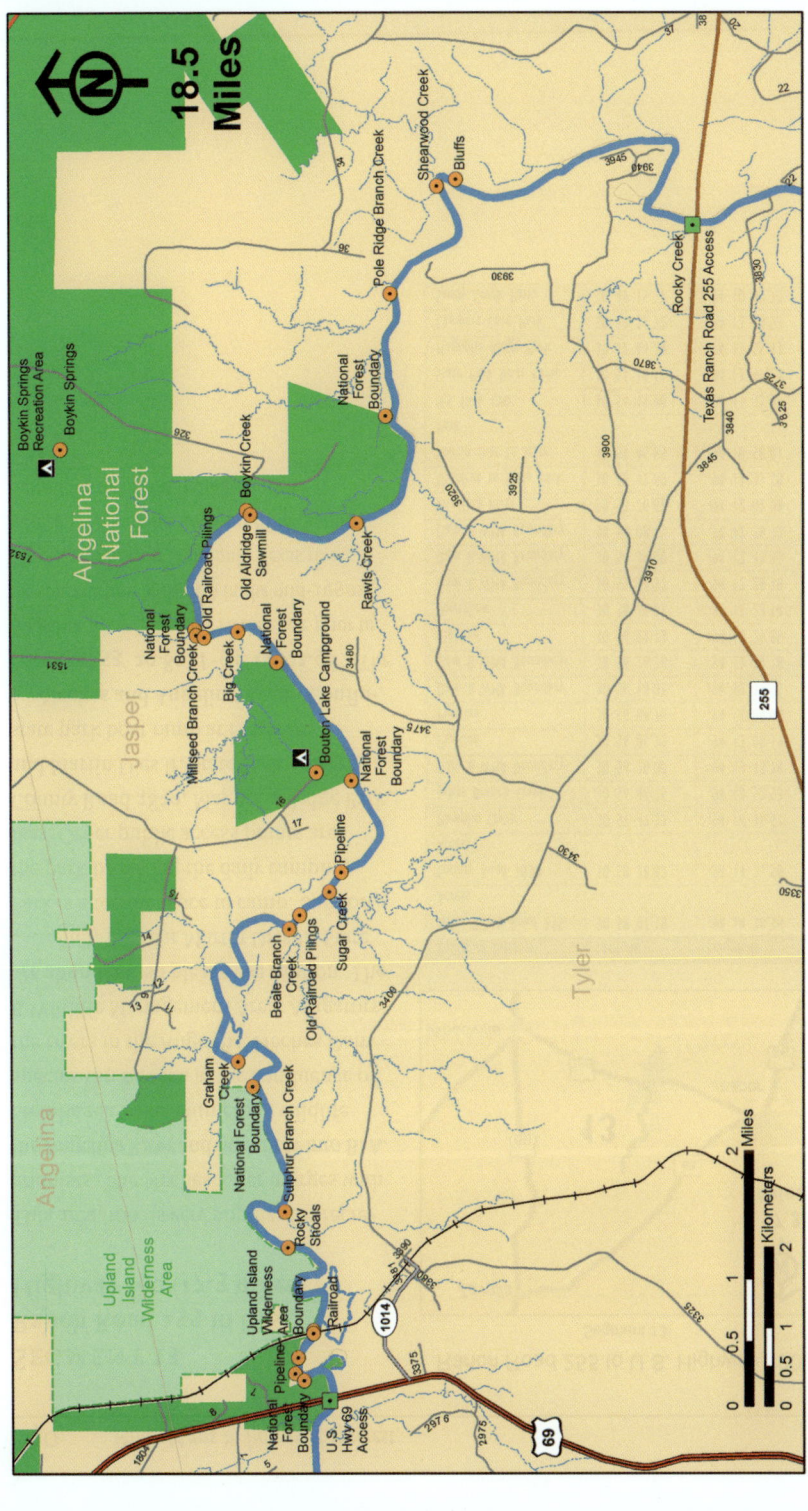

SEGMENT 13
Ranch Road 255 to U.S. Highway 190 (17.3 miles)

Ranch Road 255 to U.S. Highway 190

Segment 13

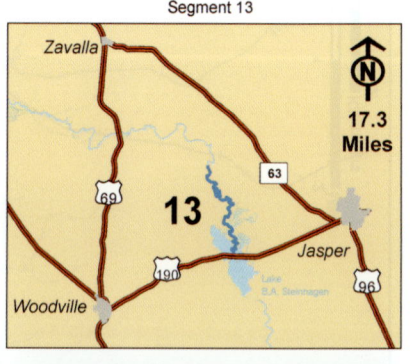

This area, too, is very popular for recreationists. The Neches River merges with the Angelina River and both flow into B. A. Steinhagen Lake. Numerous wildlife species can be seen at the confluence of the rivers in the Angelina, Neches Dam B Wildlife Management Area. Alligators are abundant, so please use caution. The extremely popular Martin Dies Jr. State Park is a terrific place to camp. Support the park by paying the daily camping fees. Other public access points are County Road 3830, Magnolia Ridge Park, and Martin Dies Jr. State Park. There is a state park boat ramp at Hwy 190.

Neches and Angelina Rivers Confluence (30° 53' 20.82" / -94° 12' 02.98") *The Angelina River joined the Neches here prior to the construction of Dam B in the mid-1950s. The Angelina River is the only Texas river named for a woman, and Angelina County is named for this same Indian maiden.*

Points of Interest	Latitude	Longitude
Texas Ranch Road 255 Access	30 59 01.18	-94 14 38.36
County Road 3830 Access	30 58 28.87	-94 14 23.58
Pamplin Creek	30 57 47.27	-94 14 36.51
Johns Branch Creek	30 56 46.75	-94 13 52.53
Dam B WMA Boundary	30 56 20.00	-94 13 13.79
Pipeline	30 56 05.23	-94 13 08.39
Pipeline	30 56 04.70	-94 13 04.08
Dam B WMA Boundary	30 56 18.01	-94 12 39.95
Dam B WMA Boundary	30 56 16.76	-94 12 30.78
Pipeline	30 56 12.13	-94 12 25.55
Powerline	30 56 11.07	-94 12 22.14
Dam B WMA Boundary	30 55 54.32	-94 12 28.38
Dam B WMA Boundary	30 54 39.00	-94 12 43.35
Dam B WMA Boundary	30 54 08.13	-94 12 58.32
Angelina River	30 53 20.82	-94 12 02.98
Shortcut to State Park	30 52 12.60	-94 11 33.58
Martin Dies Jr. State Park	30 51 45.55	-94 10 58.81
U.S. Hwy 190	30 51 11.90	-94 11 53.67
State Park Boat Ramp	30 51 16.11	-94 12 49.24
Magnolia Ridge Park	30 52 44.78	-94 13 55.43
Campers Cove Park	30 49 21.72	-94 12 04.53
Sandy Creek Park	30 49 15.77	-94 10 03.70

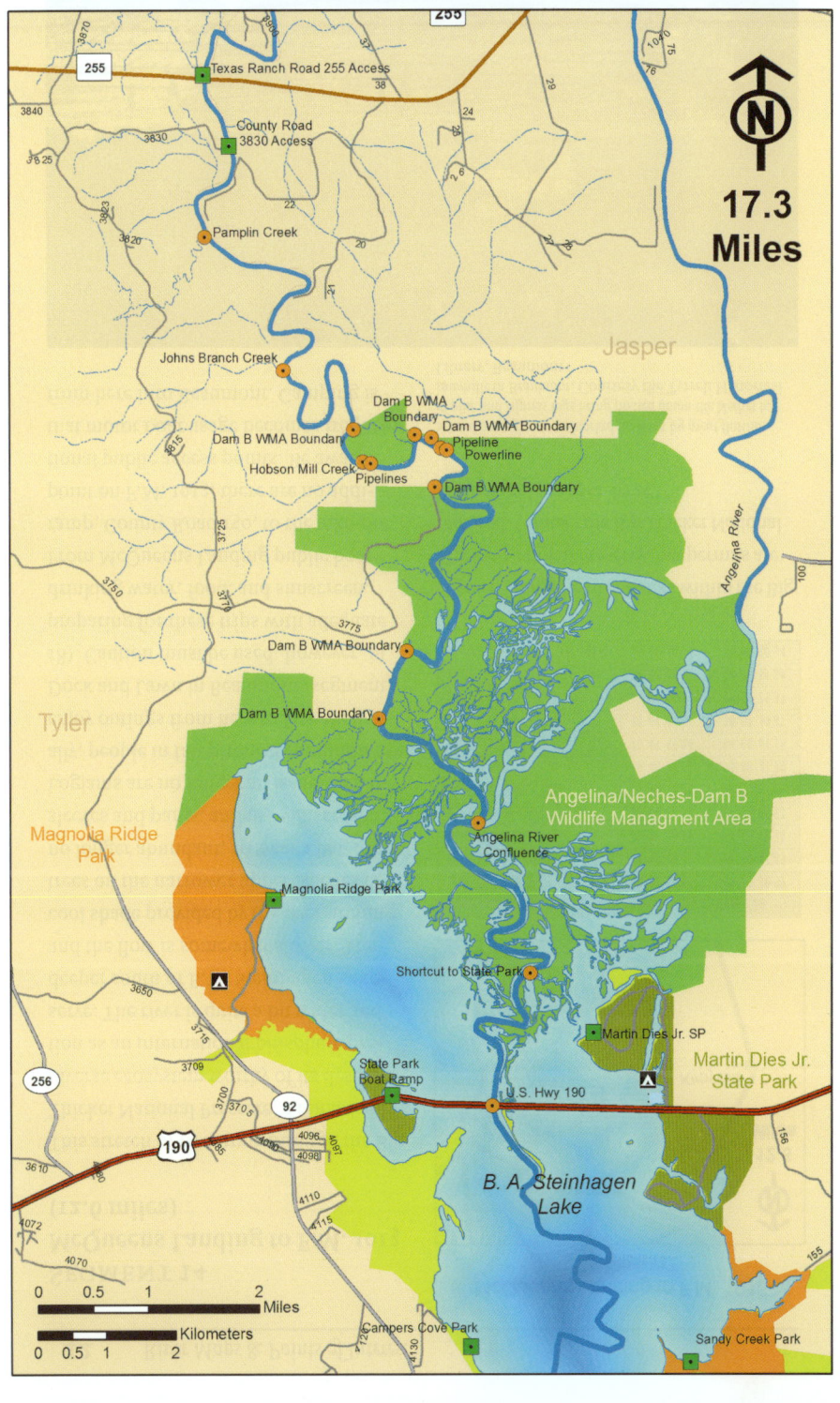

SEGMENT 14
McQueens Landing to F.M. 1013
(12.6 miles)

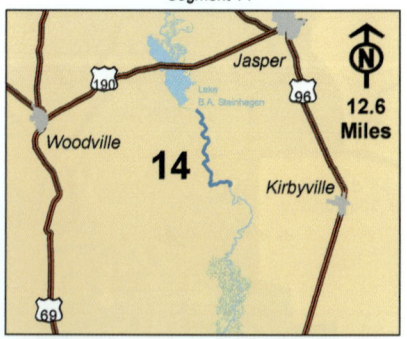

McQueens Landing to F.M. 1013
Segment 14

Jasper

Woodville

14

Kirbyville

12.6 Miles

This stretch of river runs through the Big Thicket National Preserve, a remarkably diverse ecosystem worthy of its designation as an international biosphere preserve. The river is quite a bit wider and deeper south of B. A. Steinhagen Lake, and the flow is somewhat slower. The cool shade provided by the over-arching trees on the narrower upper Neches is no longer abundant, so wear a hat, long sleeves and pants, and use sunscreen. Logjams are no longer an issue. Generally, people in fair physical condition can enjoy outings from here to Downtown Dock and Lawn in Beaumont (segment 18). Caution must be used, however, in preparing for these trips with adequate drinking water, food, and sunscreen. From McQueens Landing public boat ramp, County Road 150, to the take-out point on F.M. 1013, there are no additional public access points. Be aware that motor boat usage becomes frequent from here into Beaumont. Camping is

Points of Interest	Latitude	Longitude
Reservoir Spillway	30 47 50.67	-94 10 30.99
Reservoir Floodgates	30 47 43.42	-94 10 47.69
Outlet Channel	30 47 38.59	-94 10 13.34
Big Thicket National Preserve	30 47 36.34	-94 10 05.53
McQueens Landing Access	30 47 28.79	-94 09 03.65
Walnut Run Creek	30 46 44.58	-94 08 21.78
Big Creek	30 45 59.84	-94 08 01.15
Bluffs	30 43 44.01	-94 08 19.67
Bluffs	30 41 05.91	-94 07 35.37
Bluffs	30 41 03.57	-94 06 51.36
Sheffield Ferry Access (FM 1013)	30 40 50.79	-94 05 30.34

allowed along the river banks within the Big Thicket National Preserve, but permits are required. Contact the Big Thicket National Preserve at (409) 951-6725.

Steamboats often were blocked for days by great flotillas of pine and cypress logs being floated down the Neches to sawmills in Beaumont. Courtesy the Tyrrell Historical Library, Beaumont

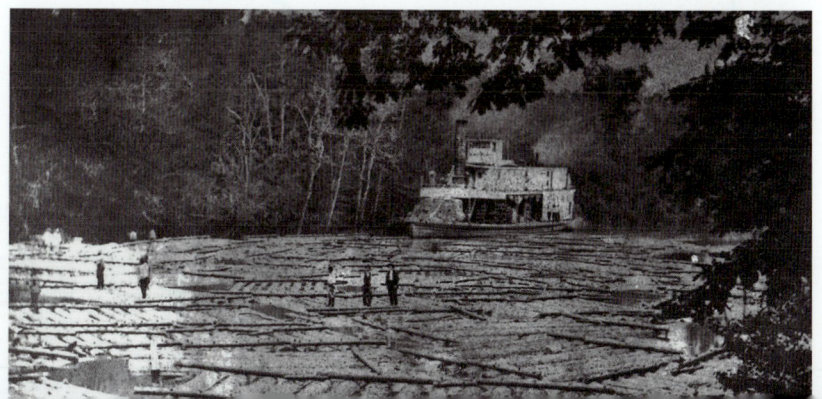

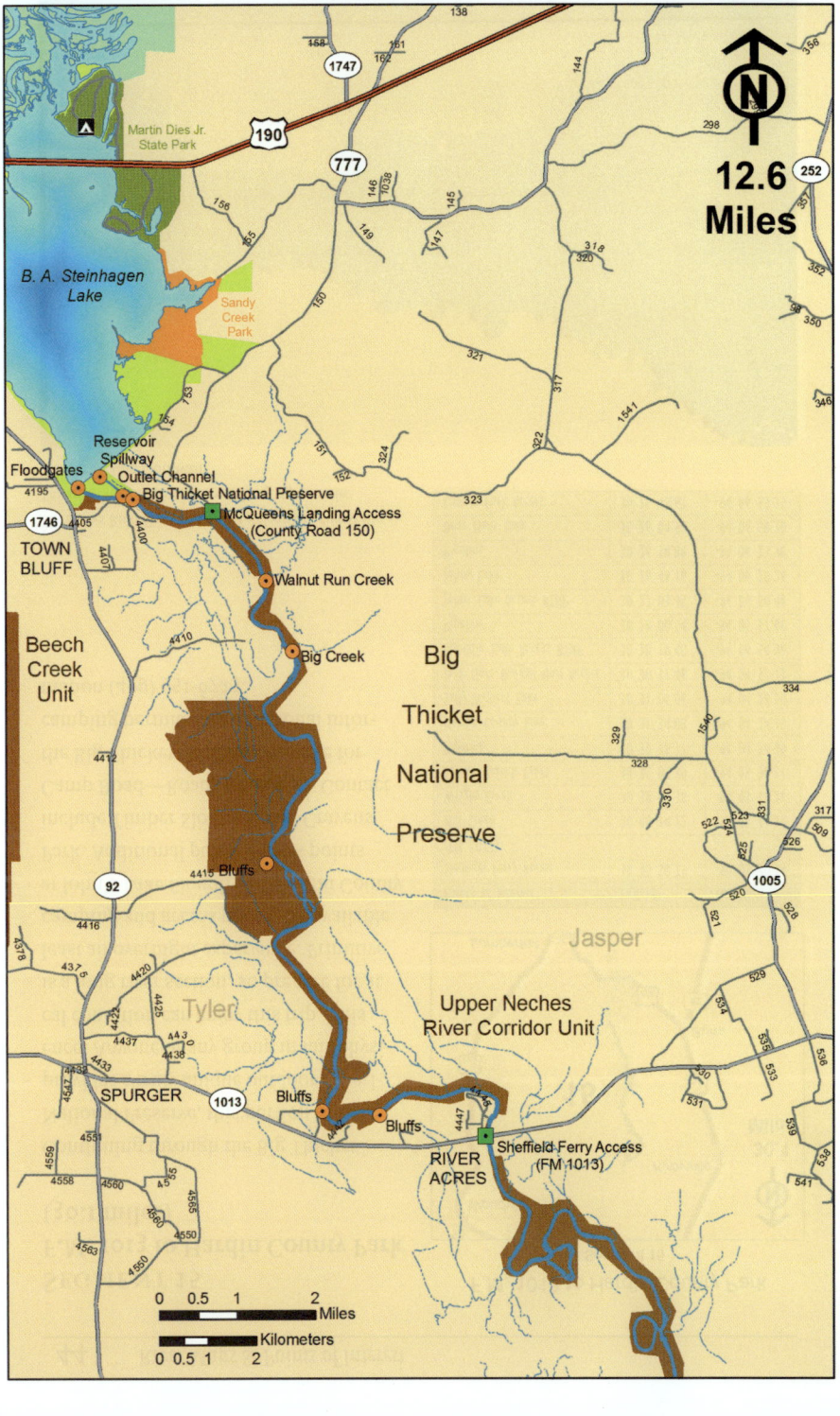

SEGMENT 15
F.M. 1013 to Hardin County Park
(30.1 miles)

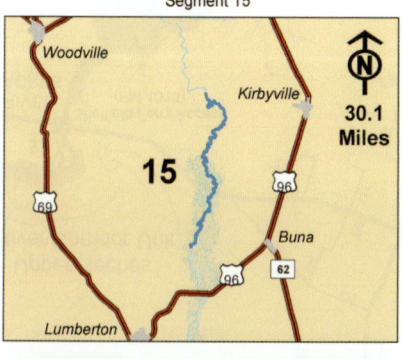

F.M. 1013 to Hardin County Park

Segment 15

Continuing through the Big Thicket National Preserve, this portion of river provides a truly unique outdoor experience. Anyone or any group in fair physical condition can enjoy this trip. This is a long river section, so prepare for at least an overnight experience. Primitive camping and access points are available at John's Lake Slough and Hardin County Park. Additional public access points include Timber Slough Road, Cravens Camp Road—Roads 95 and 96. Contact the Big Thicket National Preserve for camping permits and additional information (409) 951-6725.

Points of Interest	Latitude	Longitude
Sheffield Ferry Access (FM 1013)	30 40 50.79	-94 05 30.34
Mill Creek	30 40 14.22	-94 05 11.32
Wright Creek	30 38 44.38	-94 03 11.26
Turner Branch Creek	30 35 07.89	-94 05 30.31
Pipeline	30 32 35.77	-94 04 24.69
Hardin County Line	30 31 34.08	-94 04 24.33
Sally Withers Lake	30 31 05.06	-94 04 26.50
Jack Gore Baygall Unit Access	30 30 11.46	-94 04 51.12
Franklin Lake Access BTNP	30 28 19.40	-94 06 54.96
Pipeline	30 28 00.54	-94 05 57.60
Johns Lake Access BTNP	30 27 03.45	-94 06 50.48
Johns Lake	30 26 46.16	-94 06 29.26
Pipeline	30 26 34.88	-94 06 23.98
Bear Mans Lake	30 26 04.98	-94 06 58.20
County Park Access	30 25 43.62	-94 06 53.35

The Neches River flows though the ecologically diverse Big Thicket National Preserve for nearly eighty-six miles.
Courtesy Gina Donovan

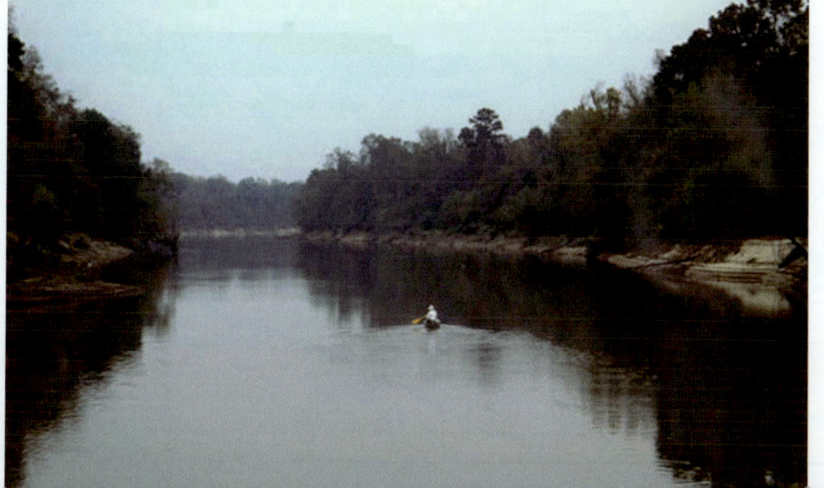

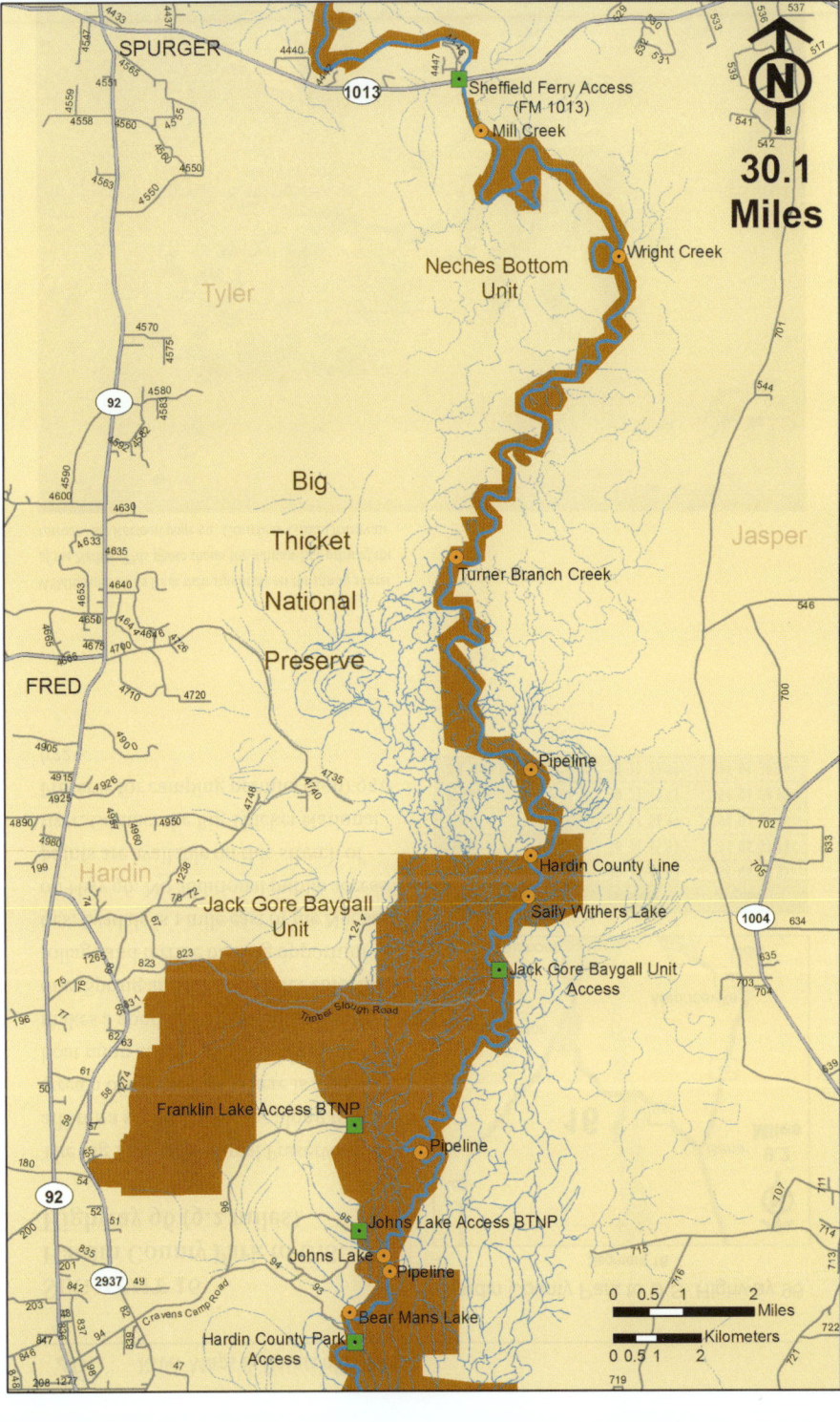

SPURGER

4433
4440
1013
Sheffield Ferry Access
(FM 1013)
Mill Creek

N

30.1
Miles

Wright Creek

Neches Bottom
Unit

Tyler

92

Big

Thicket

National

Jasper

Turner Branch Creek

Preserve

FRED

Pipeline

Hardin County Line

Hardin

Jack Gore Baygall
Unit

Sally Withers Lake

1004

Jack Gore Baygall Unit
Access

Timber Slough Road

Franklin Lake Access BTNP

Pipeline

92

Johns Lake Access BTNP

180

Johns Lake

Pipeline

2937

Cravens Camp Road

Bear Mans Lake

Hardin County Park
Access

0 0.5 1   2
Miles
Kilometers
0 0.5 1   2

719

SEGMENT 16
Hardin County Park to U.S. Highway 96 (9.2 miles)

Hardin County Park to U.S. Highway 96
Segment 16

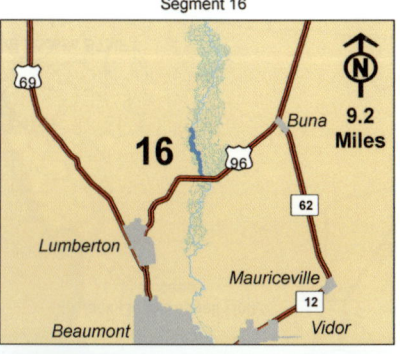

The Big Thicket National Preserve still affords a scenic landscape along this stretch of the Neches. There is a public boat ramp at Hwy 96. This segment makes a good day trip for almost everyone. Spring and fall trips offer colorful foliage and terrific birding opportunities. Evadale is 1 mile east of the Neches on Hwy 96. No additional public access points are available on this stretch of river. Contact the Big Thicket National Preserve for camping permits (409) 951-6725.

Points of Interest	Latitude	Longitude
County Park Access	30 25 43.62	-94 06 53.35
Cutoff	30 24 15.89	-94 06 57.18
Cutoff	30 24 06.99	-94 07 01.56
Pipeline	30 22 43.33	-94 06 08.90
U.S. Hwy 96 Access	30 21 22.43	-94 05 39.76

Wildlife and birds leave their signature on the sandy shores of the Neches. This photo shows the human-like print of the raccoon. See Raccoon page 59. Courtesy Gina Donovan

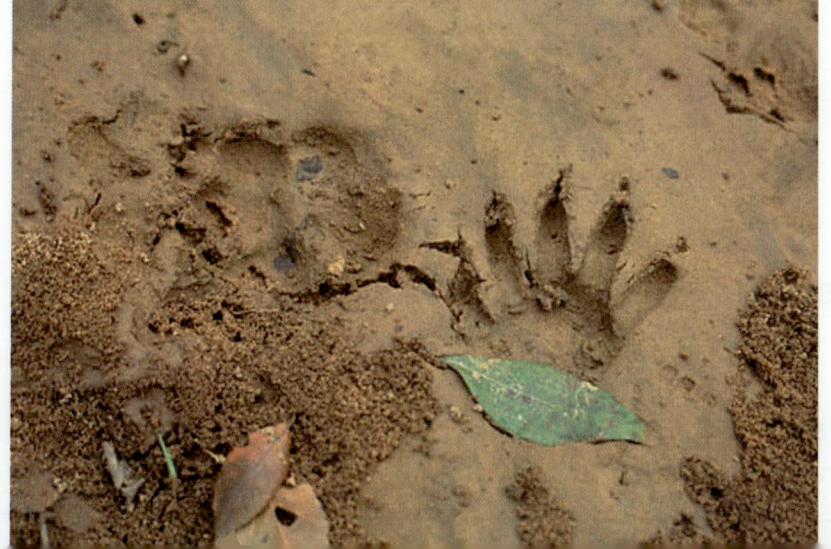

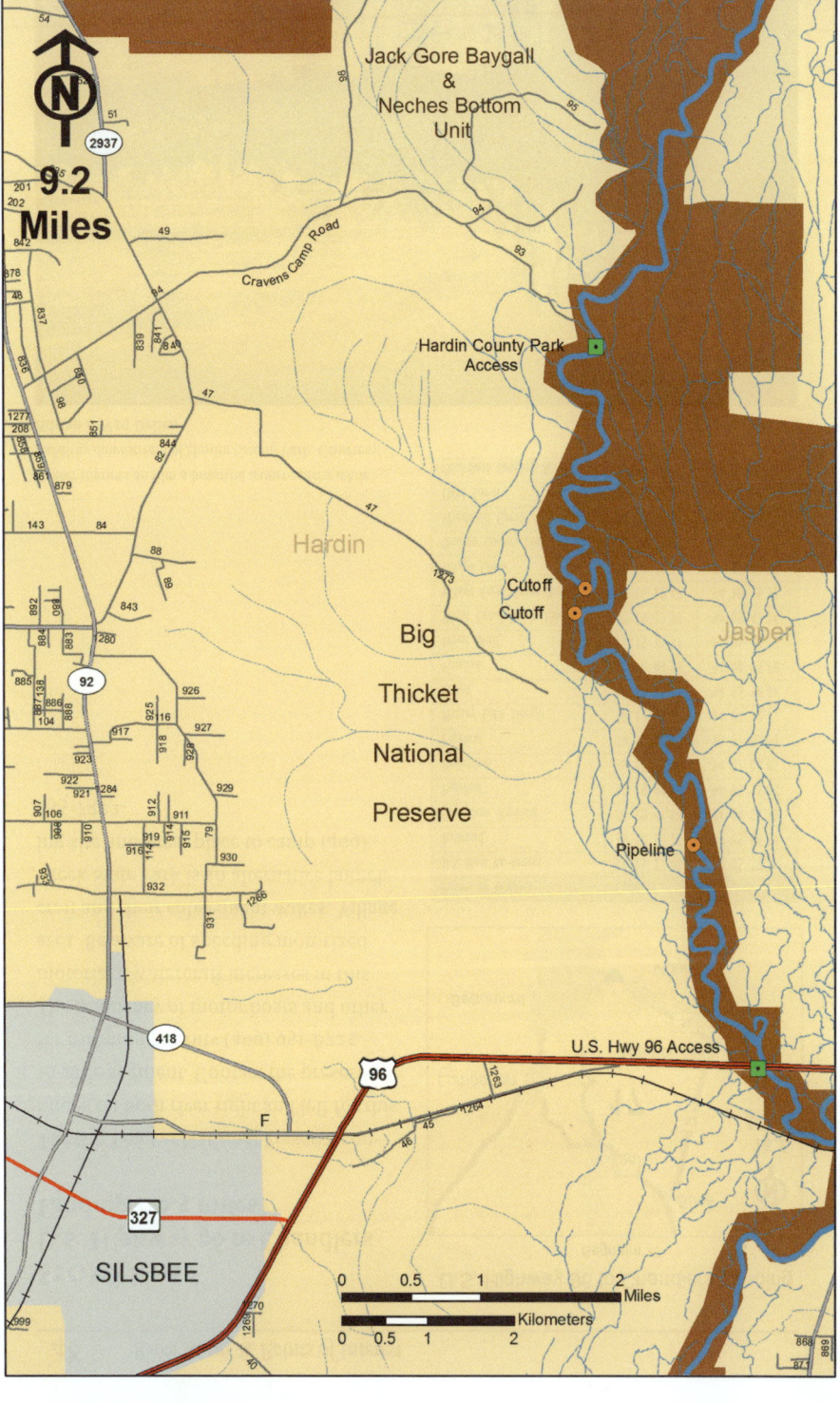

N

2937

9.2
Miles

Jack Gore Baygall
&
Neches Bottom
Unit

Cravens Camp Road

Hardin County Park
Access

Hardin

Big

Thicket

National

Preserve

Cutoff
Cutoff

Jasper

Pipeline

U.S. Hwy 96 Access

418

96

F

327

SILSBEE

| 0 | 0.5 | 1 | 2 | Miles |

Kilometers

| 0 | 0.5 | 1 | 2 |

SEGMENT 17
U.S. Highway 96 to Chandlers Landing (16.5 miles)

The Big Thicket National Preserve continues on both river right and left for this 16-mile segment. Contact the preserve for camping permits (409) 951-6725. The frequency of motor boats and other motorized watercraft increases in this area. Be aware of speeding motorized craft and their subsequent wakes. Village Creek State Park is an alternative launching site and great place to camp (409) 755-7322.

Canoer captures on film a beautiful winter sunrise while paddling downstream of Hardin County Park. Courtesy Adrian F. Van Dellen

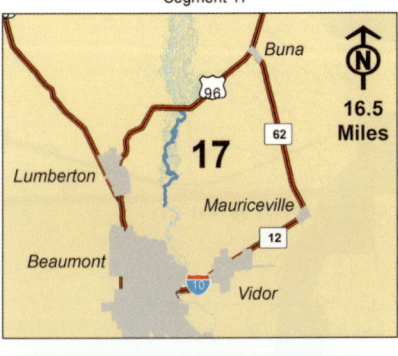

U.S. Highway 96 to Chandlers Landing
Segment 17

16.5 Miles

Points of Interest	Latitude	Longitude
U.S. Hwy 96 Access	30 21 22.43	-94 05 39.76
Railroad	30 20 53.25	-94 05 00.34
Powerline, Pipeline	30 20 41.14	-94 04 57.79
Pipeline	30 19 49.50	-94 05 55.08
Powerline	30 17 52.54	-94 07 03.45
Pipeline	30 17 48.76	-94 07 03.02
Massey Lake Slough	30 17 02.77	-94 07 12.77
Cutoff	30 15 51.18	-94 06 10.01
Pipeline	30 15 44.56	-94 06 05.76
Weiss Bluff	30 15 31.57	-94 05 45.17
Weiss Canal Pump House	30 15 22.77	-94 05 47.14
Village Creek State Park	30 15 20.17	-94 10 15.52
Village Creek	30 14 30.76	-94 07 07.89
Orange County Line	30 14 30.31	-94 07 01.35
Windfield Canal	30 13 57.67	-94 07 04.31
LNVA Canal	30 13 01.60	-94 07 04.98
Chandlers Landing Access	30 12 58.67	-94 07 00.25

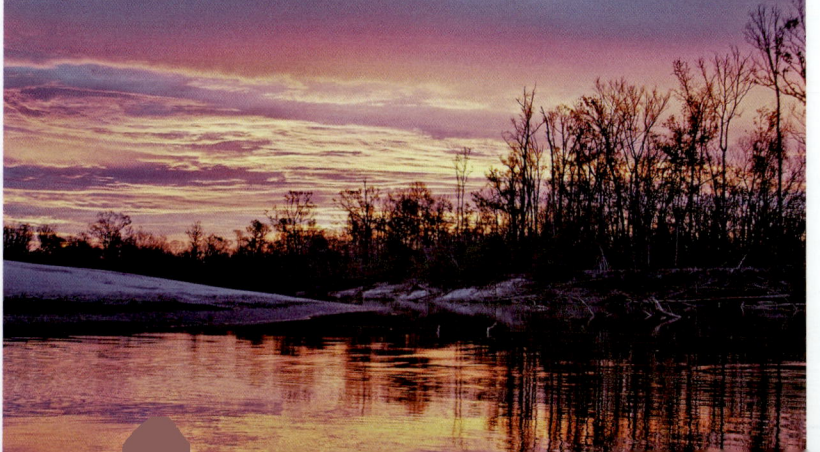

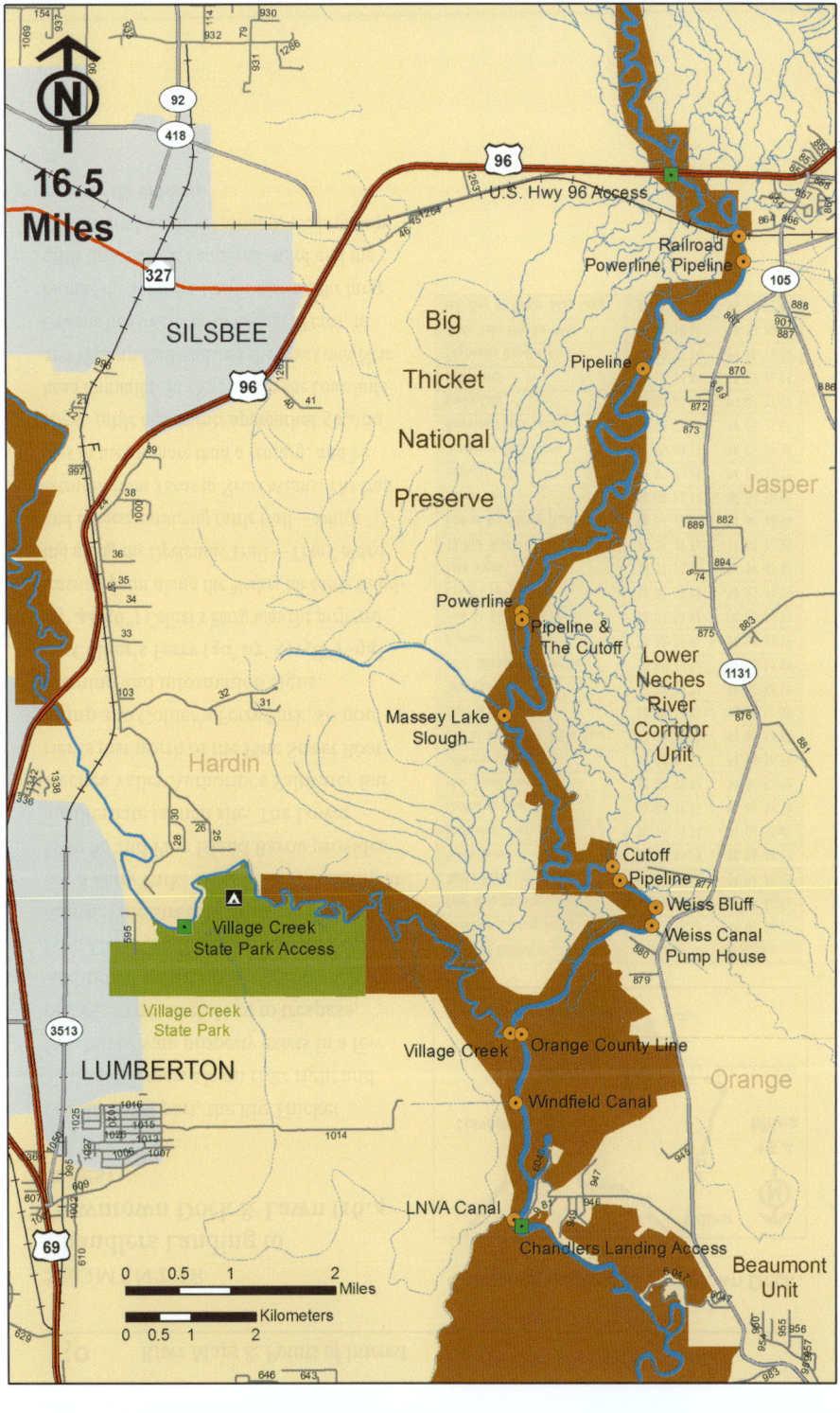

SEGMENT 18
Chandlers Landing to Downtown Dock & Lawn (16.4 miles)

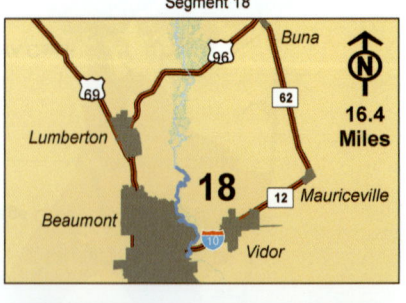

Chandlers Landing to Downtown Dock
Segment 18

16.4 Miles

For the most part, the Big Thicket National Preserve is on river right and left; but private property exists in a few places, so be careful not to trespass. Additional public access points include Four Oaks Landing, Confluence Boat Ramp, Pine Street Boat Ramp and Collier's Ferry Park. The public boat ramp at Hwy 69 and Pine Island Bayou provides an alternate launch site. The Lower Neches Valley Authority's Saltwater Barrier is just north of the Pine Street Boat Ramp and Collier's Ferry Park, so note warning and information signs.

Collier's Ferry (30° 07′ 57.58″ / -94° 05′ 44.56″) *Collier's Ferry was the preferred crossing point along the Neches for cattle traveling along the Opelousas Trail—Texas' oldest and longest-surviving cattle trail—which extended from Texas to New Orleans. The trail was in use for more than a century, and by 1855, cattle movements approached 50,000 head annually. In 1877, when the Louisiana and Western Railroad and the Texas and New Orleans line linked up at Orange, Texas, to become the Southern Pacific system, the large cattle drives across Louisiana ended and the steam train became the primary mode of shipment for the animals.*

Points of Interest	Latitude	Longitude
Chandlers Landing Access	30 12 58.67	-94 07 00.25
Four Oaks Landing	30 11 16.13	-94 05 56.61
Banks Bayou	30 10 48.64	-94 06 40.48
Sandy Lake	30 10 43.29	-94 06 44.99
Scatterman Lake	30 10 24.74	-94 06 56.81
Confluence Boat Ramp	30 09 41.75	-94 06 49.24
Pine Island Bayou	30 09 39.24	-94 06 53.02
Jefferson County Line	30 09 39.75	-94 06 56.76
End of Big Thicket (West Bank)	30 09 29.15	-94 06 54.37
Pipeline	30 09 23.85	-94 06 51.49
Saltwater Barrier Dock	30 09 19.22	-94 06 56.51
LNVA Saltwater Barrier Dam	30 09 16.54	-94 06 52.89
Pipeline	30 08 59.91	-94 06 49.16
Pine St. Boat Ramp Access	30 07 57.72	-94 05 44.72
Collier's Ferry Dock	30 07 57.58	-94 05 44.56
Lake Bayou	30 07 38.71	-94 04 43.40
10-Mile Bayou	30 06 49.76	-94 04 22.38
End of Big Thicket (East Bank)	30 06 13.31	-94 04 48.94
Bairds Bayou	30 06 12.74	-94 04 50.55
Interstate 10	30 05 40.69	-94 05 25.83
Downtown Dock Access	30 05 00.11	-94 05 39.29
Downtown Dock Lawn	30 04 56.34	-94 05 35.37
Cooks Lake	30 10 00.23	-94 07 21.57
LNVA Canal	30 10 14.84	-94 09 16.13
Edgewater Day-Use Area	30 10 39.34	-94 10 16.01
Cooks Lake Day-Use Area	30 11 18.10	-94 10 40.02
U.S. Hwy 69 Public Boat Ramp	30 10 45.72	-94 11 11.45

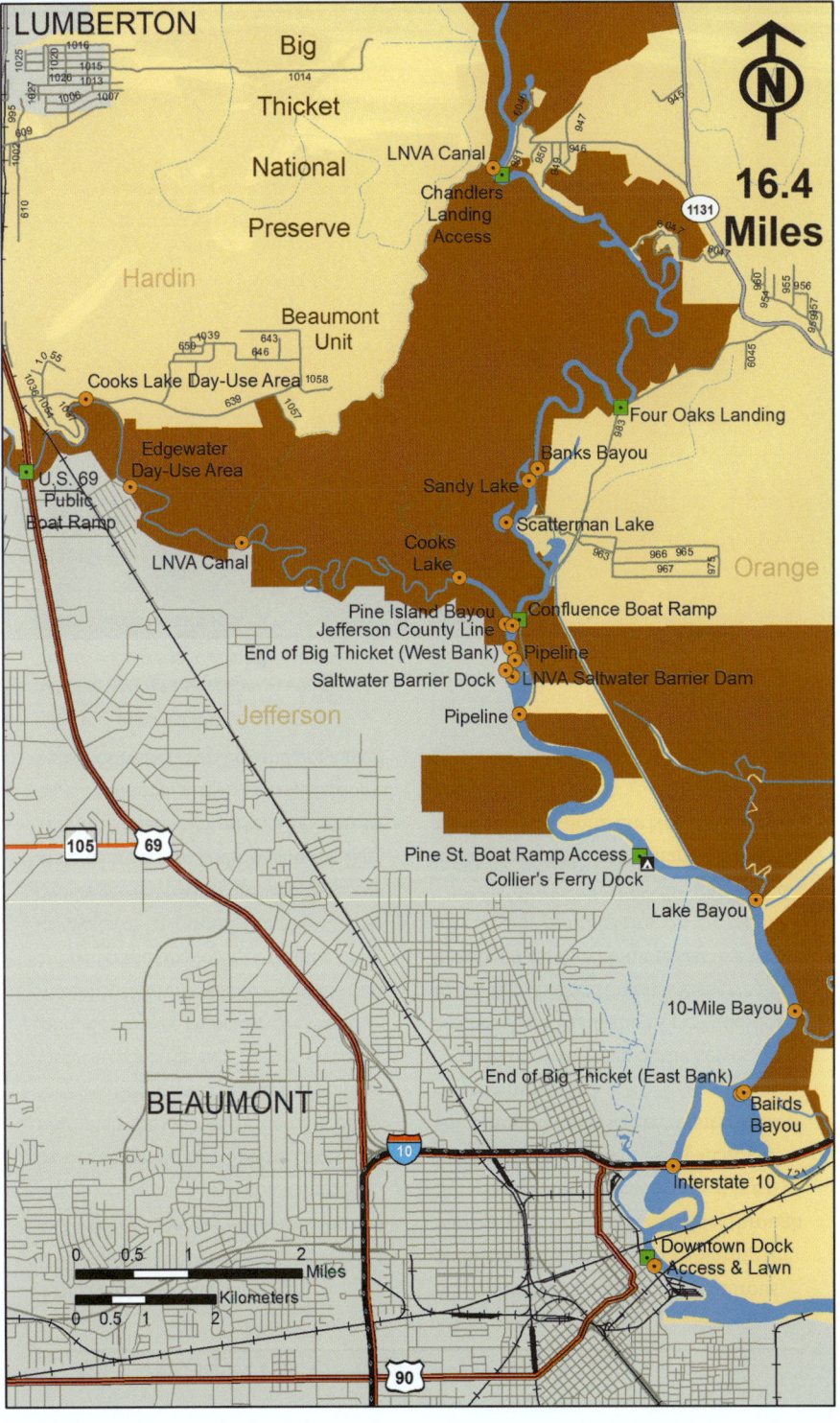

LUMBERTON

Big

Thicket

National

Preserve

Hardin

1016
1025 1020 1015
1026 1013
1097 1096 1007
995
999
1002
610

1014

LNVA Canal

Chandlers
Landing
Access

Beaumont
Unit

1039
650 643
646

Cooks Lake Day-Use Area 1058

639 1067

604B

847
946
960 861
849
607.7
6047

1131

6045

955 956
954 960
959 957

693

Four Oaks Landing

Edgewater
Day-Use Area

U.S. 69
Public
Boat Ramp

10.55
10.30 10.47
1081

LNVA Canal

Banks Bayou

Sandy Lake

Scatterman Lake

Cooks
Lake

963
966 965
967
975

Orange

Pine Island Bayou
Jefferson County Line
End of Big Thicket (West Bank)
Saltwater Barrier Dock

Confluence Boat Ramp

Pipeline

LNVA Saltwater Barrier Dam

Jefferson

Pipeline

105 69

Pine St. Boat Ramp Access
Collier's Ferry Dock

Lake Bayou

10-Mile Bayou

BEAUMONT

End of Big Thicket (East Bank)

Bairds
Bayou

10

Interstate 10

0   0.5   1   2
                        Miles

Kilometers

0   0.5 1   2

Downtown Dock
Access & Lawn

90

**16.4
Miles**

The River's End

As the river's journey ends, the bottomland forests give way to coastal marsh before emptying into Sabine Lake and ultimately the Gulf of Mexico. Before the river flows underneath Interstate 10, increasingly large expanses of swamp land, dominated by bald cypress and water tupelo, form shallow bayous that are perfect for exploring by canoe or kayak. Lake Bayou and Ten Mile Creek, two larger waterways through this ecosystem, are recent additions to the Big Thicket National Preserve. The cypress-tupelo swamps, rare in Texas, are dotted with freshwater marshes and are a mag-

net for birds and wildlife. Adventurous paddlers could spend days exploring the narrow water trails that meander through the swamps, but use caution as it is easy to become disoriented. Few patches of solid ground exist, so overnight trips are not recommended.

The following map of the Lower Neches River Basin shows new additions to the Big Thicket National Preserve that have been made within the last five years (denoted by pink shading on the map). These additions preserve the fragile and highly threatened bottomland hardwood riparian areas along the Neches River and Village Creek.

Paddling a bald cypress slough on the lower Neches in the Big Thicket National Preserve. Courtesy Adrian F. Van Dellen

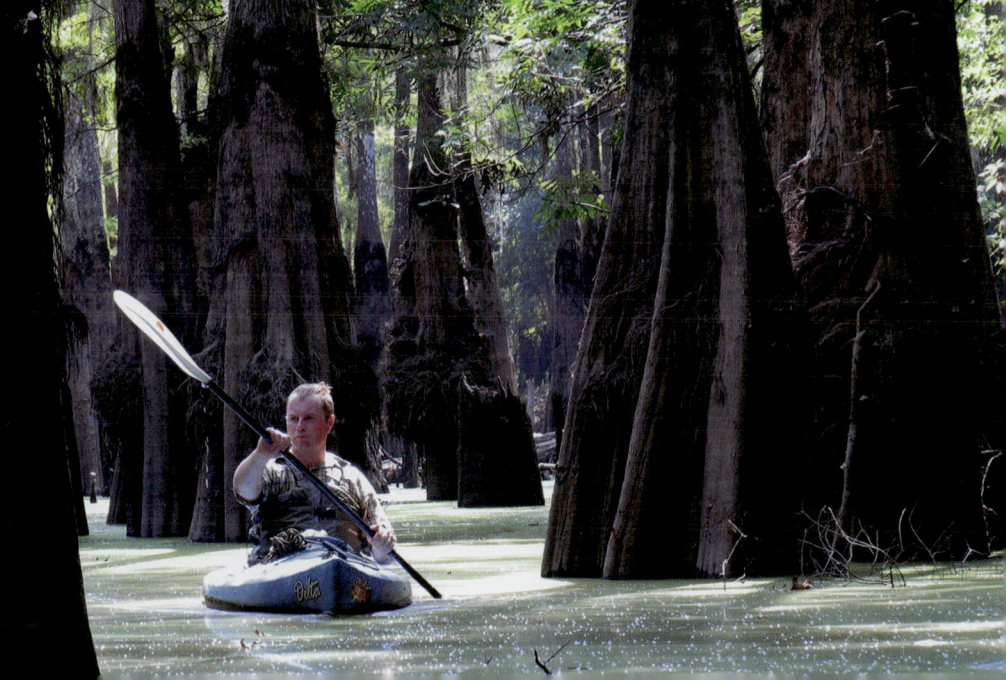

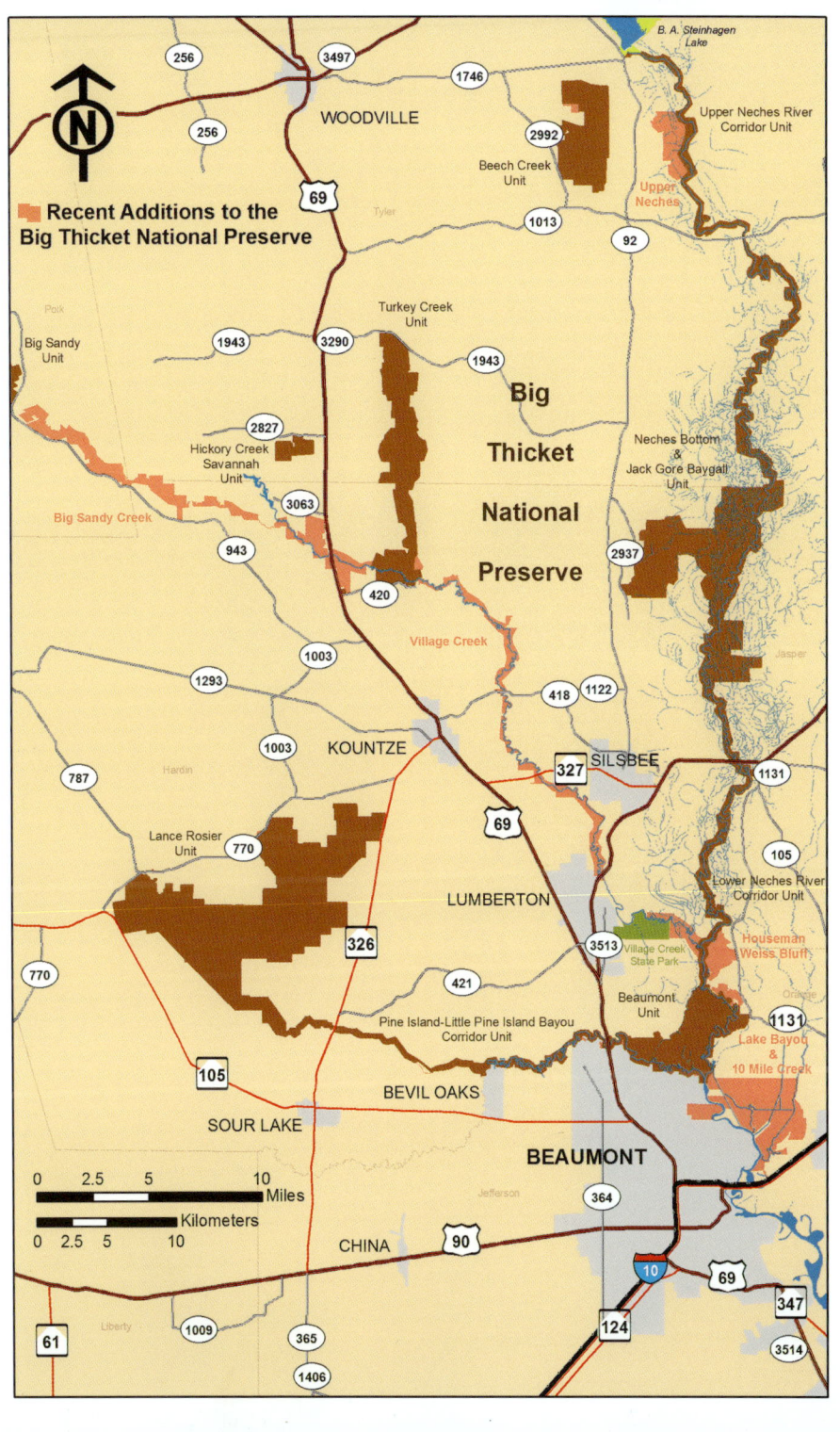

B. A. Steinhagen Lake

256   3497   1746   WOODVILLE

256   2992   Upper Neches River Corridor Unit

69   Beech Creek Unit   Upper Neches

Tyler   1013   92

Recent Additions to the Big Thicket National Preserve

Poik   Turkey Creek Unit

Big Sandy Unit   1943   3290   1943   **Big**

2827   **Thicket**

Hickory Creek Savannah Unit   Neches Bottom & Jack Gore Baygall Unit

3063   **National**

Big Sandy Creek   943   2937   **Preserve**

420   Jasper

Village Creek

1003

1293   418   1122

1003   KOUNTZE   327   SILSBEE   1131

787   Hardin   69

Lance Rosier Unit   770   105

326   Lower Neches River Corridor Unit

770   LUMBERTON   3513   Village Creek State Park   Houseman Weiss Bluff

421   Orange

Beaumont Unit   1131

Pine Island-Little Pine Island Bayou Corridor Unit   Lake Bayou & 10 Mile Creek

105   BEVIL OAKS

SOUR LAKE   BEAUMONT

0   2.5   5   10
Miles   Jefferson   364

Kilometers   10   69

0   2.5   5   10   124   347

CHINA   90

61   Liberty   1009   365   10

1406   3514

NATURAL HISTORY

"Did you see that? What in the world was it?" If you spend much time in the outdoors, there is no doubt you have said these words. The Neches River bottomlands in East Texas offer a rare opportunity to see numerous species of wildlife.

The following is not an all-inclusive list but rather a quick snapshot of what a visitor to this magnificent bottomland ecosystem may encounter.

Mammals

American Beaver (*Castor canadensis*)
A large, semi-aquatic rodent, the beaver is covered with a thick, luxurious coat of dark brown hair. It sports a large, flat paddle-shaped tail and has webbed hind feet. While despised by some people for its ability to flood acres of timberland, this animal is a big help to humans because its impoundments improve water quality and create habitat needed by numerous other animal species.

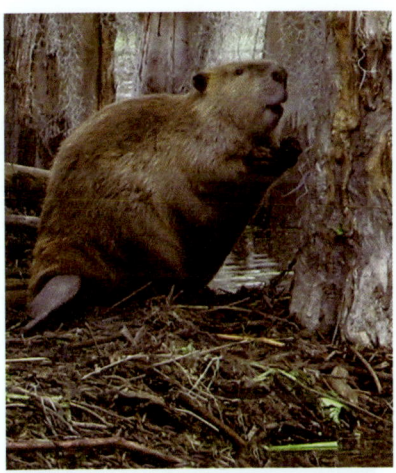

American beaver. Courtesy Anne Tindell

Bobcat (*Lynx rufus*)

A smaller cousin of the mountain lion, this cat earns its name from its bobbed tail. With colorful patterns of black and white mingled with its rust-colored fur, the bobcat has a whiskered face, black-tufted ears, and a black-tipped stubby tail. This relatively small, 35–40 pound predator makes its living off of smaller mammals, primarily rabbits, rodents, and some birds.

Coyote (*Canis latrans*)

A coyote has grayish brown to yellowish brown fur, with whitish fur on its belly and a long bushy tail. Weighing from 20–50 pounds, the coyote is sometimes mistaken for a medium-sized dog. More often heard than seen, these animals hunt primarily at night and prefer to avoid humans. Many travelers in the river bottom will hear the evening yips, barks, and howls of a coyote. People sometimes think coyotes are wolves; however, wolves are actually extinct in East Texas.

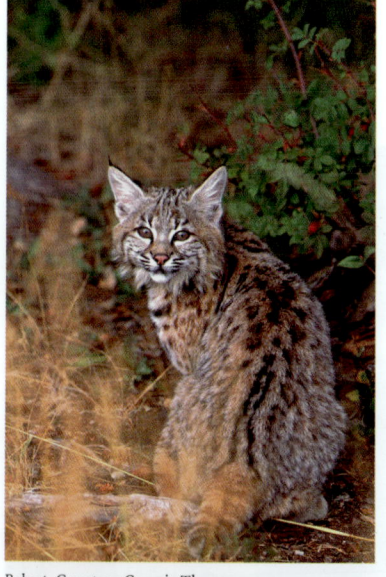

Bobcat. Courtesy Connie Thompson

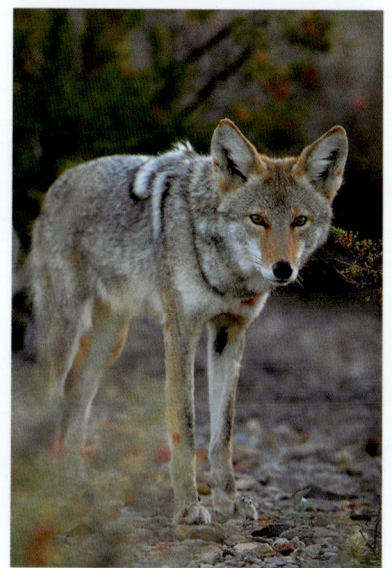

Coyote. Courtesy Greg Lavaty

Eastern Cottontail Rabbit (*Sylvilagus floridanus*)

The eastern cottontail is red-brown or gray-brown in appearance and has large hind feet, long ears, and a short fluffy white tail, which gives it its name. Usually nesting in burrows, this rabbit prefers more upland sites along the fringes of wet areas.

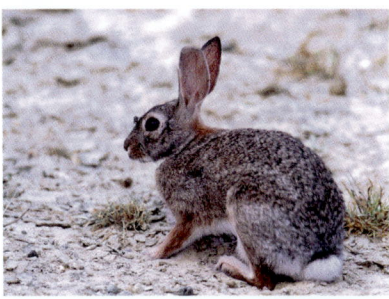

Eastern cottontail rabbit. Courtesy Connie Thompson

Eastern Fox Squirrel (*Sciurus niger*)

The fox squirrel is the largest species of tree squirrel native to North America. Weighing up to 3 pounds, the fox squirrel is the color of a red fox, with variations from orange to yellowish, and has a bushy tail. In the wild, fox squirrels primarily depend on tree seeds for food, but they are generalist eaters and will also consume buds, fruits, insects, bird eggs, and small lizards.

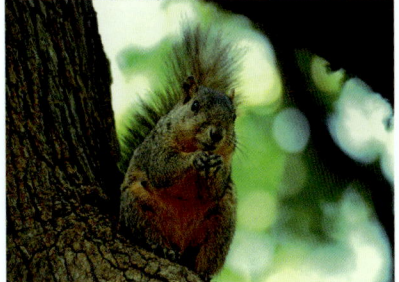

Eastern fox squirrel. Courtesy Connie Thompson

Eastern Gray Squirrel (*Sciurus carolinensis*)

As its name suggests, the eastern gray squirrel's fur is predominantly gray, but it occasionally sports a reddish tinge. The animal's bushy tail has white-tipped hairs. This squirrel is sometimes called the "farmer" of the forest due to its habit of burying nuts and acorns for later use (many of which are never recovered and later sprout into new trees). The gray squirrel will spend the majority of its life in the trees and will not venture far from one in case danger arises. Smaller than the fox squirrel, the gray squirrel weighs about a pound.

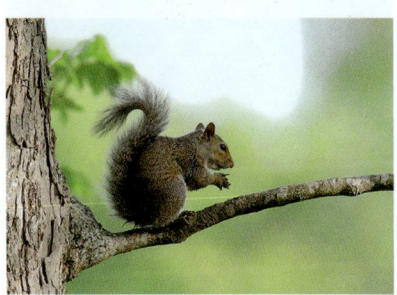

Eastern gray squirrel. Courtesy Greg Lavaty

Feral Hog (*Sus scrofa*)

Feral or wild hogs look a lot like domestic hogs and may vary in coat color and pattern. Black, brown, white, and any combination of the three is possible. Weighing up to 400 pounds, this animal may be active night or day. Males are generally larger than females. Feral hogs can cause a lot of damage to ecosystems. They compete for food with many native animals and, in fact, consider some of them prey. They also can cause considerable damage to native plant communities and cause soil erosion.

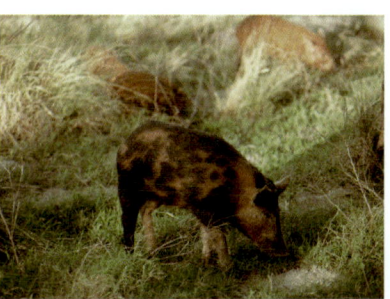

Feral hog. Courtesy Texas Parks and Wildlife Department

Mountain Lion (*Felis concolor*)

If you are lucky enough to actually see one of these shy, reclusive creatures, you are one of the elite few. Hunting and habitat changes have caused the mountain lion to be one of the very rare finds in the Pineywoods. Frequently called a cougar, panther, or puma, the mountain lion is the only large long-tailed cat found in this area. The mountain lion's tawny or tan color is typical of the species. A huge cat may weigh up to 200 pounds, but 75– 140 pounds is more typical.

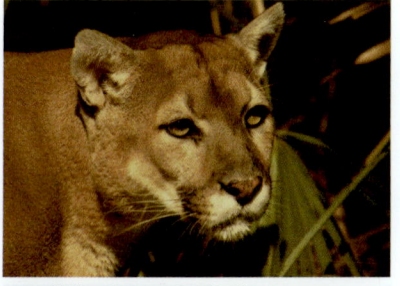

Mountain lion. Courtesy Connie Thompson

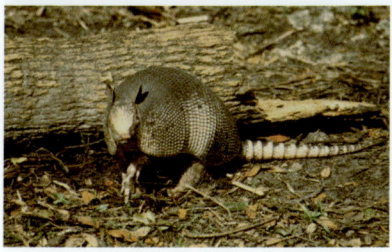

Nine-banded armadillo. Courtesy Connie Thompson

Nine-Banded Armadillo (*Dasypus novemcinctus*)

Another unusual mammal, the armadillo carries its own suit of armor for protection. Typically a gray color, the upper portion of its body is covered with bony plates and sparse, coarse hairs cover its belly. Eating primarily insects and worms, it roots and digs throughout the night in search of food. Its activities help the forest floor by aerating and turning the soil, exposing seed sources. This species is the only known animal able to inflate its intestines to float across a river.

Raccoon (Procyon lotor)

Sometimes called the masked bandit, the raccoon has gray to brown fur, a white face covered with a black mask, and a bushy, ringed tail. The raccoon's paws look like miniature human hands, making the animal easily identifiable along the sandy banks of the Neches River. Usually, raccoons are nocturnal (active at night).

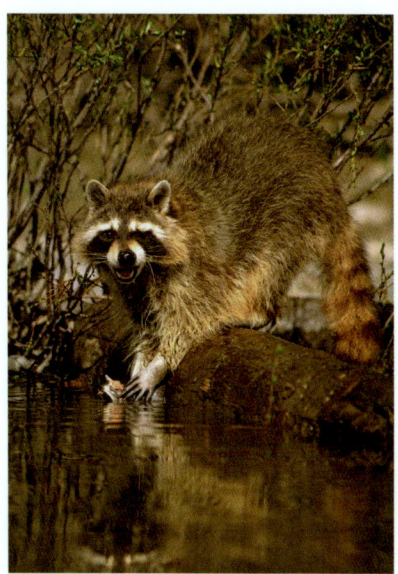

Raccoon. Courtesy Connie Thompson

Red Fox (Vulpes vulpes)

The red fox is not native to Texas but was introduced here in the late 1800s for sport hunting purposes. The red fox is primarily a rusty red color with a white underbelly, black ear tips and legs, and a bushy tail with a white tip. This animal prefers heavily wooded habitats and river bottomlands. The red fox is an opportunistic feeder but is primarily carnivorous. It prefers small mammals but will also consume birds, eggs, insects, and some berries and fruits. The red fox is often mistaken for the gray fox, a Texas native that is not included in this guide because it prefers more upland habitat.

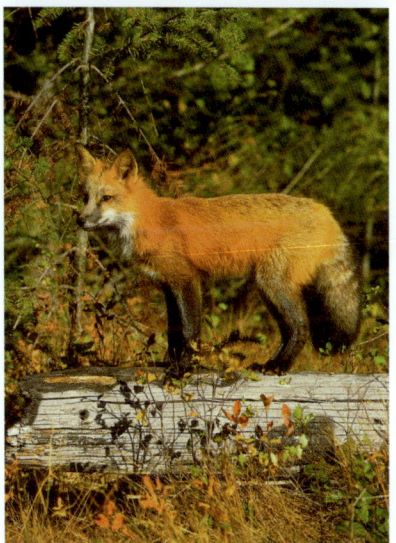

Red fox. Courtesy Connie Thompson

River Otter (*Lutra canadensis*)
Weighing 15–25 pounds, river otters have velvety thick fur ranging in color from nearly black to reddish or grayish brown. River otters are expert swimmers and divers and dwell in the river itself; however, they frequent the shore to forage, play, and groom. Eating crayfish, fish, small turtles, and other aquatic prey, river otters are one of the more playful and intelligent inhabitants of the river bottom. They are most active from early evening until early morning.

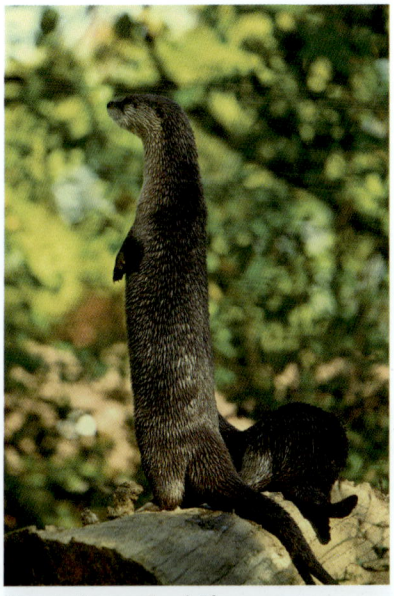

River otter. Courtesy Connie Thompson

Striped Skunk (*Mephitis mephitis*)
The striped skunk, black in color, can be identified by the white stripes that run from its head to its tail. Its stripes start with a triangle at the head and break into two stripes down its back, which meet again to form one stripe at the base of the tail. The striped skunk is about the size of a house cat, and is primarily nocturnal. The skunk is not a favorite animal of most people because of the strong-smelling liquid it will spray up to 12 feet as a defense measure to deter predators (including humans).

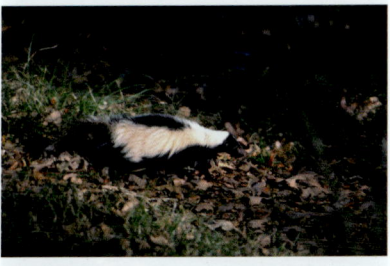

Striped skunk. Courtesy Connie Thompson

Swamp Rabbit (*Sylvilagus aquaticus*)
Similar in appearance to the cottontail, although larger, the swamp rabbit is not shy about getting into the water. Swamp rabbits have rusty colored feet and usually nest beneath logs in marshy and swampy habitats.

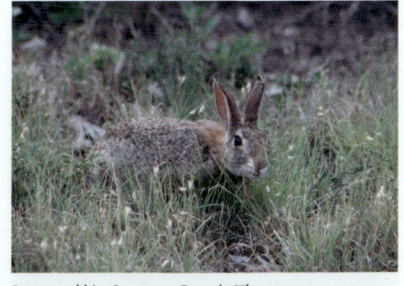

Swamp rabbit. Courtesy Connie Thompson

Virginia Opossum (*Didelphis virginiana*)

The Virginia opossum is North America's only marsupial, which means it carries its babies in a pouch on its belly like a kangaroo or koala. The opossum has a coat of gray fur and a long naked tail. It is nocturnal and uses its keen sense of smell to locate food. The opossum is about the size of a large house cat.

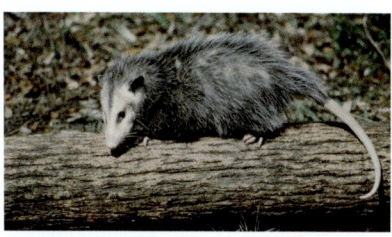

Virginia opossum. Courtesy Connie Thompson

White-Tailed Deer (*Odocoileus virginianus*)

The white-tailed deer is a tan or reddish brown color in the summer and a grayish brown in winter. It has white on its belly, throat, and the underside of its tail. When startled, the white-tailed deer may "flag" or raise its tail and show its white underside. Fawns are reddish brown at birth with white spots. Male deer have antlers.

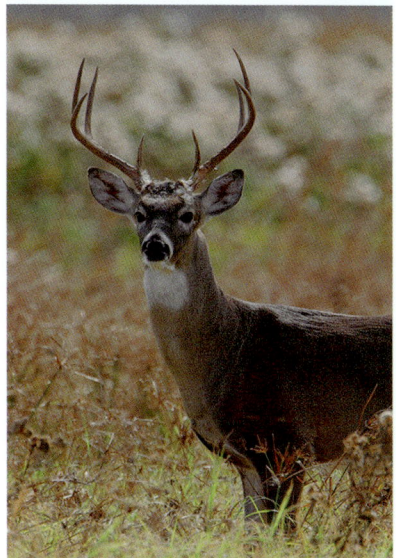

White-tailed deer. Courtesy Greg Lavaty

Reptiles and Amphibians

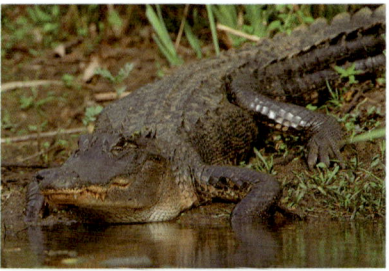

American alligator. Courtesy Greg Lavaty

American Alligator (*Alligator mississippiensis*)

Almost black in color, the American alligator is a large, semi-aquatic armored reptile and can range from 6 to 14 feet long. It has prominent eyes and nostrils and coarse scales over its entire body. An agile swimmer, the American alligator often floats or swims with only its eyes and nostrils exposed. Alligators generally avoid humans, but use caution when around this reptile.

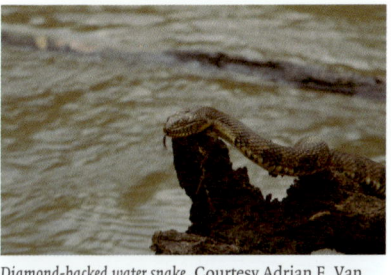

Diamond-backed water snake. Courtesy Adrian F. Van Dellen

Diamond-Backed Water Snake (*Nerodia rhombifer rhombifer*)

Often confused with the western cottonmouth, the diamond-backed water snake has a more visible skin pattern. Blackish brown lines form a diamond-shaped network on its grayish brown body. This aquatic snake lacks venom and will typically flee human intrusion. During the hot summer months, the snake is usually active only at night. (Nonvenomous)

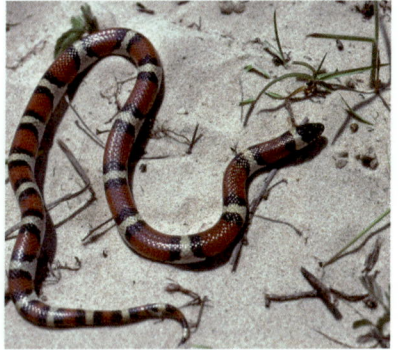

Louisiana milk snake. Courtesy Texas Parks and Wildlife Department

Louisiana Milk Snake (*Lampropeltis triangulum amaura*)

This brightly colored snake is one of nature's "mimic" animals because it looks very similar to the highly venomous Texas coral snake. Although nonvenomous, the Louisiana milk snake also sports the black, red, and yellow bands, but the red and yellow bands do not touch. The old saying, "red and black venom lack," is a good way to differentiate between the Texas coral snake and the Louisiana milk snake. (Nonvenomous)

Red-Eared Slider (*Trachemys scripta elegans*)

Red-eared sliders are Texas' most common aquatic turtles and are perhaps the most recognizable. These medium-sized turtles have a dark green oval shell and a red stripe behind their eyes. These turtles can be observed sliding off rocks and logs when startled.

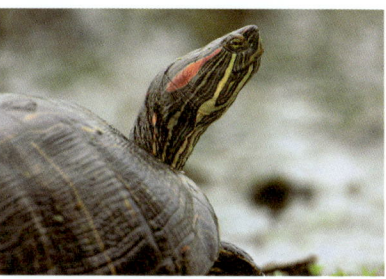

Red-eared slider. Courtesy Greg Lavaty

Southern Copperhead (*Agkistrodon contortrix contortrix*)

Copperheads in this area tend to have a pale brown to light tan body with a pinkish tint and a copper-colored pattern. When seen from the side, its markings look a lot like Hershey's Kisses; but when seen from above, its pattern resembles an hour glass. Copperheads prefer swampy habitats and eat small rodents and frogs. (Venomous)

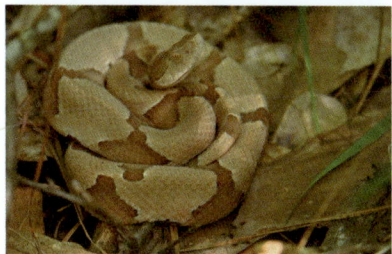

Southern copperhead. Courtesy Adrian F. Van Dellen

Texas Coral Snake (*Micrurus fulvius tener*)

The old axiom, "red and yellow kill a fellow," applies to this colorful, but secretive reptile. The snake's distinctive pattern is a broad black band, a narrow yellow band, and a broad red band, with the red and yellow bands touching. Although typically not aggressive toward humans, the Texas coral snake is one of the most virulently toxined snakes in North America, so use caution when in the presence of one. The snake's diet consists primarily of lizards and other small snakes. (Venomous)

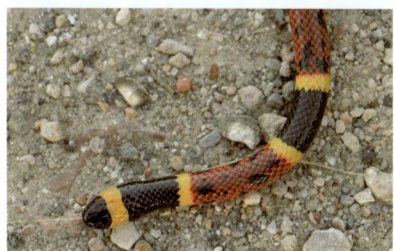

Texas coral snake. Courtesy Greg Lavaty

Texas Rat Snake (*Elaphe obsoleta lindheimeri*)

Also known as the "chicken snake," the Texas rat snake comes in a variety of colors ranging from gray to tan to brown to black. It's most often recognized by its pattern of large, dark brown rectangular blotches on tan-colored skin. The rat snake lives in a variety of habitats and has a wide ranging diet. It has earned a bad reputation as the snake most often responsible for raiding chicken coops. When threatened, this snake sometimes tries to fool predators by "rattling" its naked tail in the leaves to imitate a rattlesnake. (Nonvenomous)

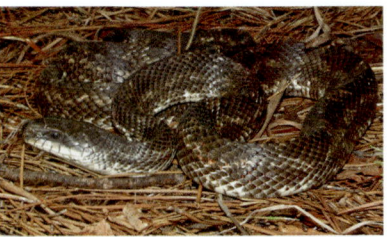

Texas rat snake. Courtesy Brendan Kavanagh

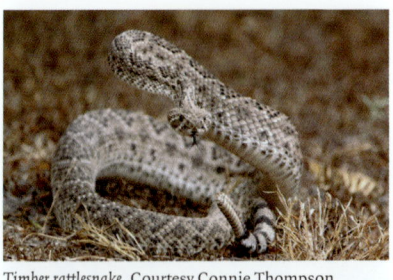

Timber rattlesnake. Courtesy Connie Thompson

Timber Rattlesnake (*Crotalus horridus*)

A threatened species in Texas, the timber rattlesnake, sometimes called the canebrake rattler, is identifiable by its grayish color and black bands, and by its entirely black tail, rattles, and cinnamon-hued vertebral stripe. Typically, this snake is not very aggressive and prefers to be left alone to hunt small mammals and hide in the leaf litter of the forest floor. Found primarily in cane breaks in moist river corridors, the timber rattlesnake rarely grows over 60 inches. (Venomous)

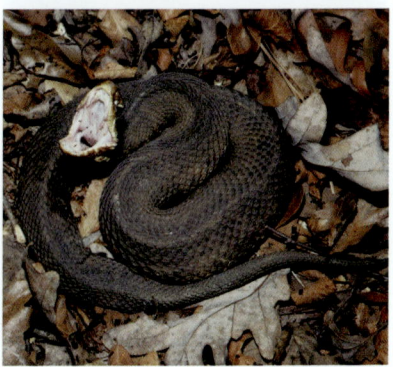

Western cottonmouth. Courtesy Craig Rudolph

Western Cottonmouth (*Agkistrodon piscivorus leucostoma*)

With little to no visible pattern or variation to its dark color, this stout thick-bodied snake lives primarily in the water but will sun and sometimes hunt on the banks and in the branches above water.

When frightened, the cottonmouth will pop its mouth open revealing the bright white skin inside, hence the name "cottonmouth." Cottonmouths are not picky eaters and will often stand their ground with humans, only giving in reluctantly. (Venomous)

Birds

Bald Eagle (*Haliaeetus leucocephalus*)
The bald eagle is one of nature's most impressive birds of prey. These birds have a white head, neck, and tail and a dark brown body. They can weigh up to 10 pounds and generally have a 6- to 7-foot wingspan. Bald eagles are commonly seen in East Texas in November through March. The eagles are usually spotted perched in tall trees or soaring overhead. (Migrant—winter)

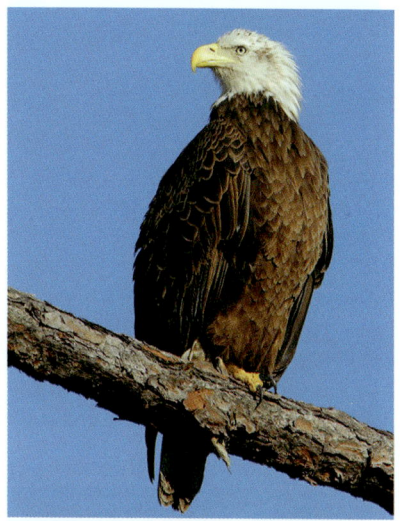
Bald eagle. Courtesy Greg Lavaty

Barred Owl (*Strix varia*)
A relatively stocky, medium-sized owl with dark eyes, the barred owl has a pale-colored breast marked with brown streaks. Generally nocturnal, these owls have a distinctive call which sounds a lot like "who cooks for you, who cooks for you all?" They prefer hardwood swamps. (Resident)

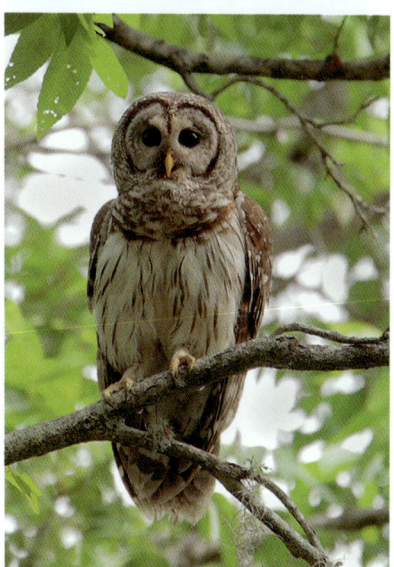
Barred owl. Courtesy Greg Lavaty

Belted Kingfisher (*Ceryle alcyon*)
Belted kingfishers are often seen perched on wires or tree branches close to water. They hover over water to locate small fish, their primary food source, before diving headfirst after their prey. Male belted kingfishers have a slate blue breast band, and the female has an added rust belly band and flanks. Their distinctive loud, rattling call is easy to recognize. Kingfishers dig burrows in river banks for their nests. (Resident)

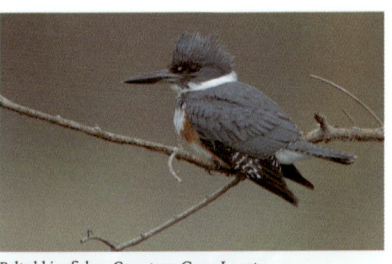

Belted kingfisher. Courtesy Greg Lavaty

Black Vulture (*Coragyps atratus*)
Black vultures may be recognized by their black bodies and the white patches at the end of their wings. These birds have bald, dark-colored heads. They are usually seen soaring overhead looking for carrion (dead animals). Large communal roosts may also be spotted from the river. (Resident)

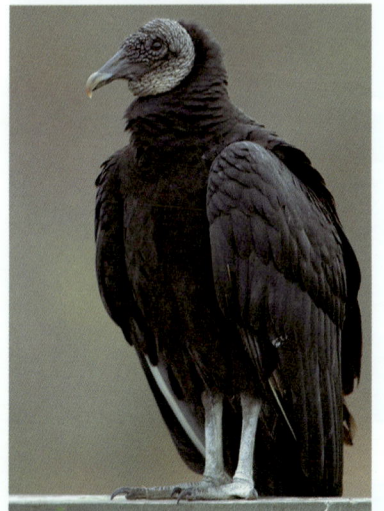

Black vulture. Courtesy Greg Lavaty

Great Blue Heron (*Ardea herodias*)
The great blue heron is one of our largest birds and is gray to blue-gray in color. This bird has a strong, thick yellow bill and dark legs. It is typically solitary and can be found along the banks of the Neches hunting for fish, frogs, small mammals, and other prey. (Resident)

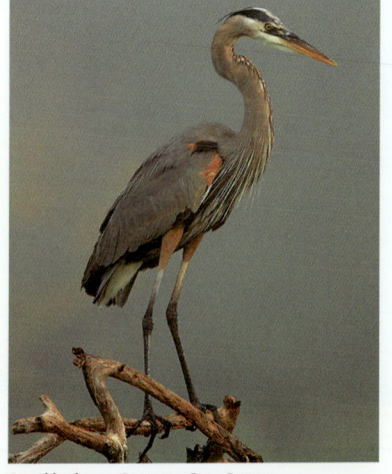

Great blue heron. Courtesy Greg Lavaty

Great Egret (*Ardea alba*)
Texas' largest white heron and common along the Neches, the great egret has a long neck, a thick yellow bill, and black legs. These birds are usually seen on tree snags or flying along the river's corridor. Great egrets nest in large communal roosts. (Resident)

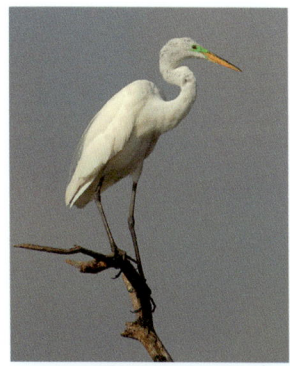

Great egret. Courtesy Greg Lavaty

Great Horned Owl (*Bubo virginianus*)
The largest owl in Texas, great horned owls have prominent "ear" tufts and bright yellow eyes. They prefer upland habitats and are chiefly nocturnal. These owls can take prey as large as a skunk. Their call is a rhythmic series of three to eight loud, deep hoots. (Resident)

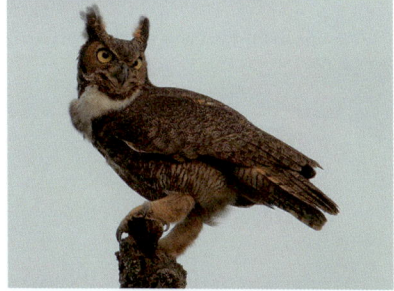

Great horned owl. Courtesy Greg Lavaty

Northern Parula (*Parula americana*)
The northern parula is a tiny, short-tailed warbler with gray-blue above and a yellowish green tint on its upper back, two white wing bars, and a yellow throat and chest. Its song is a rising trill ending with a sharp zip. These birds are found along the Neches only during the summer, as they are a migratory species. (Migrant—summer)

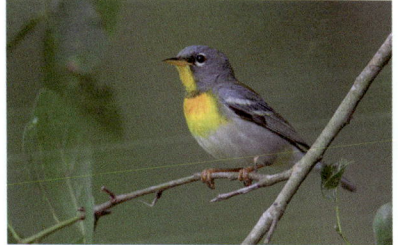

Northern parula. Courtesy Greg Lavaty

Pileated Woodpecker (*Dryocopus pileatus*)
The pileated woodpecker is a large, crow-sized woodpecker, with a brilliant red crest and white wing patches. People often mistake this bird for the ivory-billed woodpecker, thought by some to be extinct in North America. The pileated woodpecker's preferred food source is carpenter ants, which are usually found

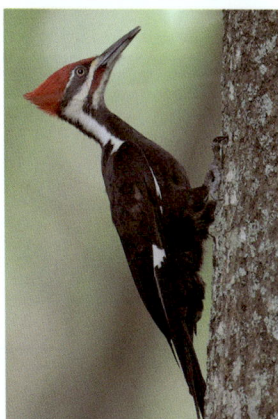

Pileated woodpecker. Courtesy Greg Lavaty

in tall dead trees along the river; these trees also make ideal nesting sites. The bird's loud, haunting call is often heard along the Neches. (Resident)

Prothonotary Warbler (*Protonotaria citrea*)

A migratory species, prothonotary warblers are found along the Neches during the summer. They are brilliant yellow with blue-gray wings. (Migrant—summer)

Scarlet Tanager (*Piranga olivacea*)

The scarlet tanager only passes through East Texas on its way north to its nesting grounds; however, this bird uses the abundant deciduous forests along the Neches River for resting and refueling on insects and larvae before continuing its journey. The male scarlet tanager is an unmistakable scarlet or red color with black wings and tail. The females are a greenish yellow color with dark or black wings. (Migrant)

Summer Tanager (*Piranga rubra*)

The male summer tanager is a bright, rosy red, and the female is a yellow-olive hue with a reddish wash. These birds are common in mature woods, especially bottomland hardwood forests. (Migrant—summer)

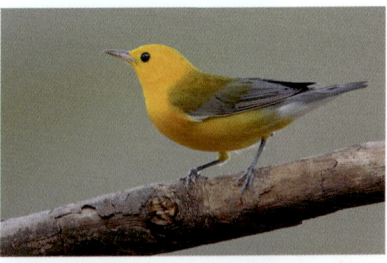

Prothonotary warbler. Courtesy Greg Lavaty

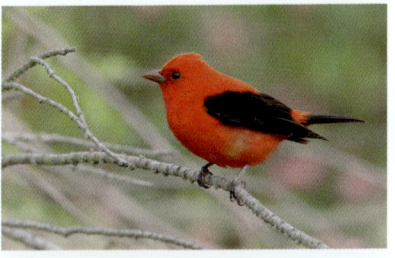

Scarlet tanager. Courtesy Greg Lavaty

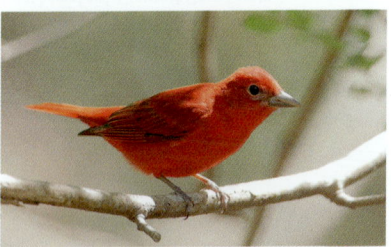

Summer tanager (male). Courtesy Greg Lavaty

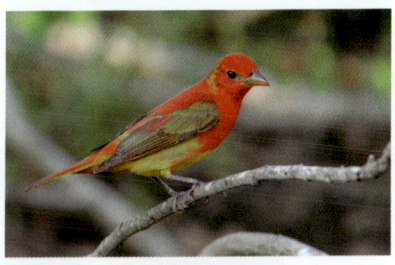

Summer tanager (female). Courtesy Greg Lavaty

Turkey Vulture *(Cathartes aura)*

Turkey vultures are very large birds, almost as big as bald eagles. They appear black or brown with bald, red heads. Like black vultures, turkey vultures are usually seen circling overhead searching for carrion. (Resident)

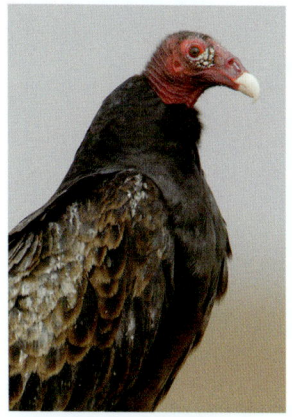

Turkey vulture. Courtesy Greg Lavaty

White-Eyed Vireo *(Vireo griseus)*

The white-eyed vireo can be found year-round on the lower stretches of the Neches, but only on the river's upper reaches during the summer. This bird is a grayish olive above and white below, with pale yellow sides and flanks and two white wing bars. This thicket-loving bird is difficult to see but is often heard. Its typical song is a loud five to seven note phrase usually beginning and ending with a sharp "chick." (Resident to summer migrant)

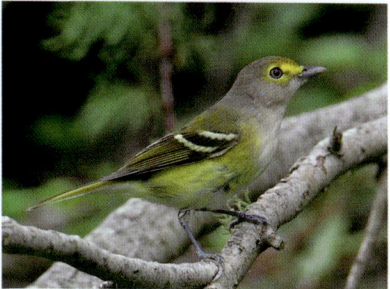

White-eyed vireo. Courtesy Greg Lavaty

Wood Duck *(Aix sponsa)*

Wood ducks are generally seen along the Neches when flushed from their resting or feeding areas among fallen trees in the river. Males have glossy, brilliantly colored feathers, and females are a muted grayish green. These ducks nest in tree cavities, and the young leave the nest at a very early age and may be seen swimming behind their mother. (Resident)

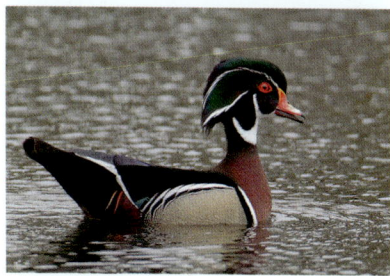

Wood duck. Courtesy Greg Lavaty

Yellow-Billed Cuckoo (*Coccyzus americanus*)

The yellow-billed cuckoo is a summer visitor to the Neches River Valley. It is grayish brown above, white below, and has large white spots on the underside of its dark-colored tail. It feeds on caterpillars and other prey gleaned from tree leaves and branches. (Migrant—summer)

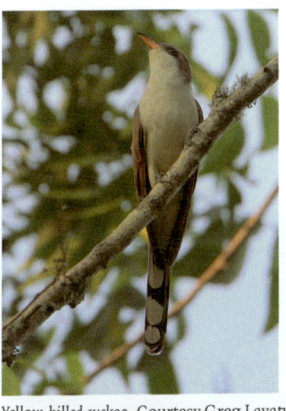

Yellow-billed cuckoo. Courtesy Greg Lavaty

Trees and Vegetation

American Holly (*Ilex opaca*)

This evergreen, spiny leafed tree has red berries in the fall that provides food for songbirds, deer, and wild turkeys during the winter. Many local people use the holly's evergreen branches and red berries as Christmas decorations.

Bald Cypress (*Taxodium distichum*)

Bald cypress trees grow in swamps and in wet bottomlands along rivers and streams. This large deciduous conifer is known for its prominent knees that project from the ground. The tree's seeds are eaten by a number of bird species, including wild ducks.

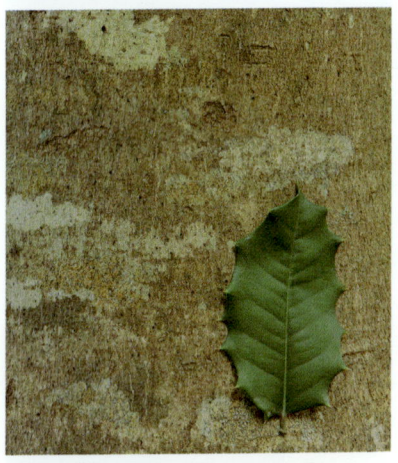

American holly. Courtesy Adrian F. Van Dellen

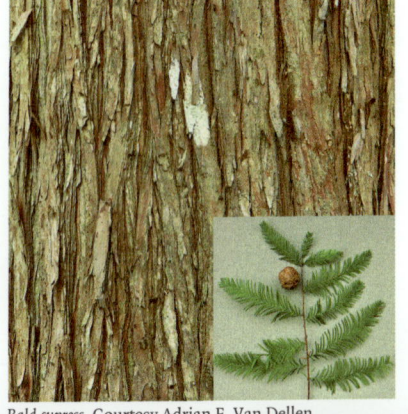

Bald cypress. Courtesy Adrian F. Van Dellen

Black Tupelo (*Nyssa sylvatica*)
Known locally as "black gum," the black tupelo is famous for its brilliant red fall color and is easily seen along the Neches. The tree's fruit attracts and feeds thirty-two bird species, and its leaves are browsed by white-tailed deer and black bear.

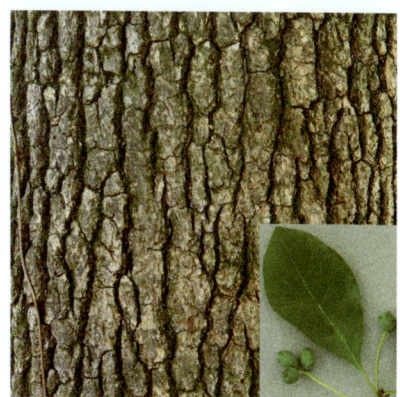

Black tupelo. Courtesy Adrian F. Van Dellen

Black Willow (*Salix nigra*)
The black willow is a medium-sized deciduous tree that often grows along the edge of water with narrow leaves and fluffy white seeds. The black willow's leaves are green on both sides.

Chinese Tallow-Tree (*Sapium sebiferum*)
The Chinese tallow is a nonnative tree species from China that has spread over hundreds of thousands of acres in the United States, crowding out native, wildlife-friendly vegetation. It is a medium-sized deciduous tree known for its beautiful fall colors of yellow, red, and orange. Its leaves are diamond-shaped.

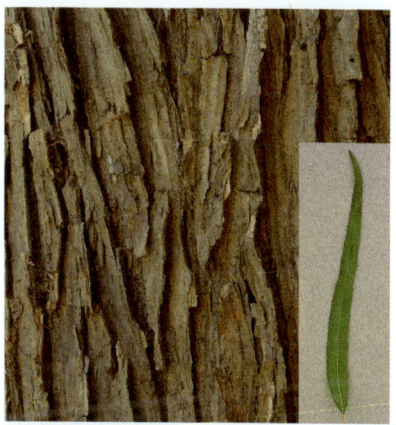

Black willow. Courtesy Adrian F. Van Dellen

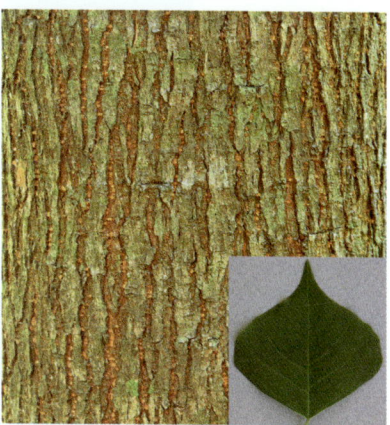

Chinese tallow-tree. Courtesy Adrian F. Van Dellen

Cocklebur (*Xanthium strumarium*)

Many nature lovers have encountered the infamous cocklebur. This plant produces hundreds of miniature football-shaped burs that stick to animal fur and human clothing. Many fingers have been pricked while trying to remove scores of the prickly burs from shoestrings, socks, and pants after a walk along the river bank.

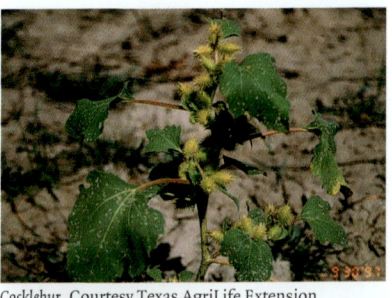

Cocklebur. Courtesy Texas AgriLife Extension

Common Buttonbush (*Cephalanthus occidentalis*)

The buttonbush grows best in moist soils and has white, round flowers that attract butterflies during the summer. During September and October, the buttonbush produces round clusters of reddish brown nutlets, which are eaten by at least twenty-five bird species, mostly water birds. The shrub also provides good bee food.

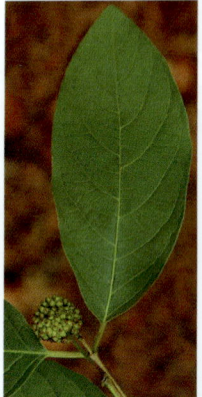

Common Buttonbush. Courtesy Adrian F. Van Dellen

Dwarf Palmetto (*Sabal minor*)

The dwarf palmetto is a trunkless palm that often grows in wetland areas with large, fanlike evergreen leaves. It is one of only two native Texas palm species.

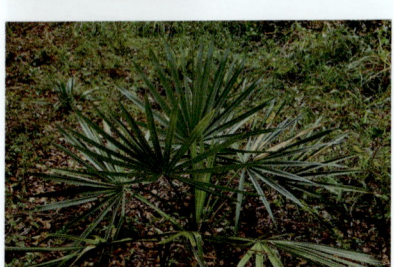

Dwarf palmetto. Courtesy Adrian F. Van Dellen

Eastern Persimmon (*Diospyros virginiana*)

The eastern persimmon is a medium-sized deciduous tree that produces a tasty orange-colored fruit in late fall. Humans and many bird species, opossums, raccoons, skunks, squirrels, foxes, deer, and feral hogs enjoy eating this sweet fruit, especially after the first frost.

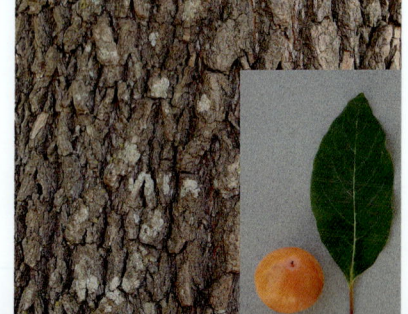

Giant Cane (*Arundinaria gigantea*)

Known in East Texas as "switch cane," giant cane is a woody grass that grows to about 23 feet in height. Giant cane is America's only native bamboo, and was used by American Indians and early settlers alike to make spears, arrows, blowguns, pipes, flutes, and fish traps. This woody grass once formed large, impenetrable cane brakes across the South.

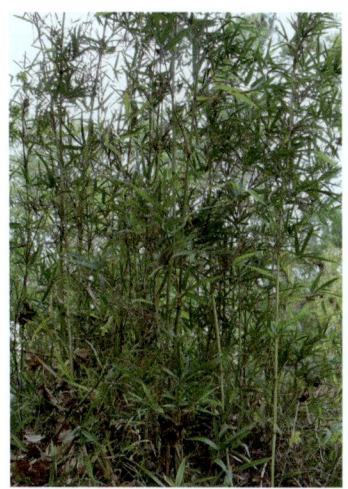

Giant cane. Courtesy Adrian F. Van Dellen

Loblolly Pine (*Pinus taeda*)

This fast-growing yellow pine is abundant in East Texas and can be identified by its needle-like leaves and its dark-colored deeply furrowed bark. The loblolly pine is an evergreen tree with brown seed-bearing cones.

Mimosa (*Albizia julibrissin*)

The mimosa is a fine textured, medium-sized nonnative tree from China with fragrant, pink "powder puff" flowers in the spring. Sometimes called "silk-tree," it is an invasive, exotic species throughout the South.

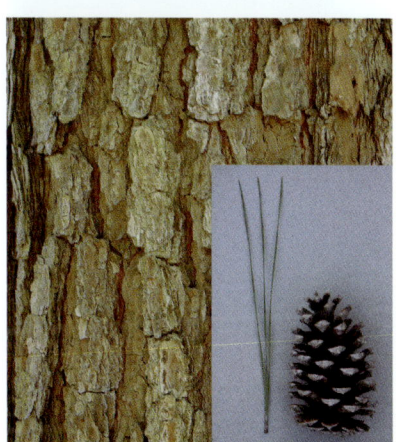

Loblolly pine. Courtesy Adrian F. Van Dellen

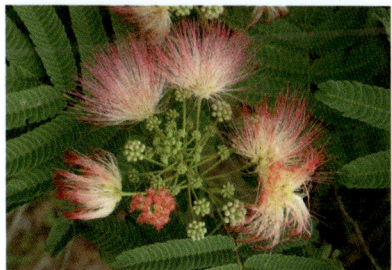

Mimosa. Courtesy Adrian F. Van Dellen

Overcup Oak (*Quercus lyrata*)

The acorns of an overcup oak are easily identifiable by the "cup" that covers almost all the round, somewhat flattened acorn. The tree's leaves are long, 7–9 inches, and are a variety of shapes. Its grayish bark is rough and flaky. Deer and cattle browse on young overcup oak trees.

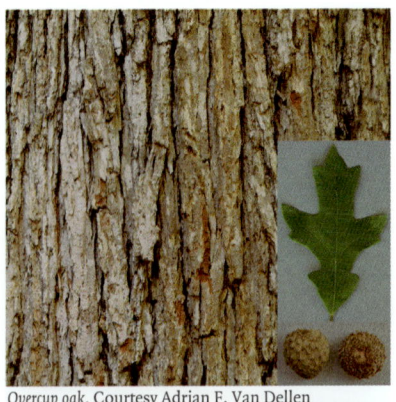

Overcup oak. Courtesy Adrian F. Van Dellen

Planer Tree (*Planera aquatica*)

Commonly known as the "water elm," the planer tree is native to North America and is found in abundance along the upper portions of the Neches River. This small spreading tree grows to about 40 feet in height. Its bark is scaly, patchy, and grayish brown. Its fruit is considered to be an important duck food. Squirrels also eat the tree's fruit.

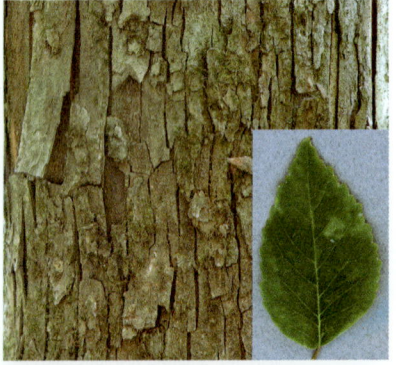

Planer tree. Courtesy Adrian F. Van Dellen

River Birch (*Betula nigra*)

The unique attribute of the river birch is its tough, papery, peeling bark. The bark ranges in color from reddish brown to cinnamon-red, and the peeling layers appear ragged and quite distinctive. The river birch is native to East Texas and has arrow-shaped leaves.

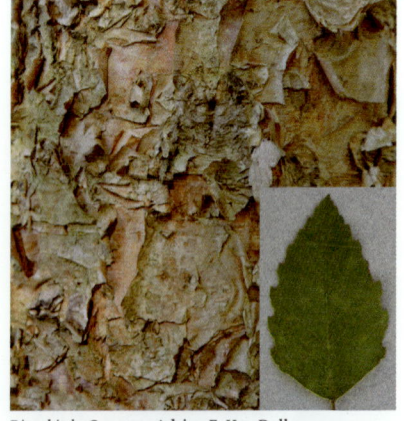

River birch. Courtesy Adrian F. Van Dellen

Southern Wood Fern (*Thelypteris kunthii*)

The wood fern is a deciduous, terrestrial fern that has lance-shaped fronds. It is generally found in moist shady areas.

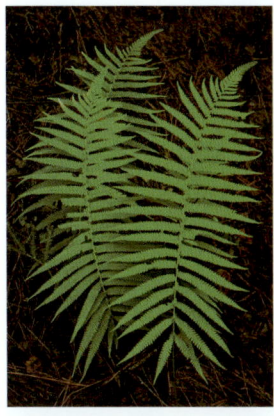

Southern wood fern. Courtesy Adrian F. Van Dellen

Swamp Chestnut Oak (*Quercus michauxii*)

Swamp chestnut oak leaves are green above, paler and pubescent or furry beneath, 4–8 inches in length, and turn a crimson color in the fall. Its acorns are eaten by mourning doves, wild turkeys, and white-tailed deer. Its leaves are frequently browsed by livestock. The tree's wood is used for posts, tools, baskets, boards, veneer, and fuel.

Sweetgum (*Liquidambar styraciflua*)

The sweetgum is known for its star-shaped leaves, attractive fall color, and prickly "gum ball" fruit. Its bark is light gray and has corky scales. At least twenty-five bird species feed upon the tree's fruit, as does the eastern gray squirrel.

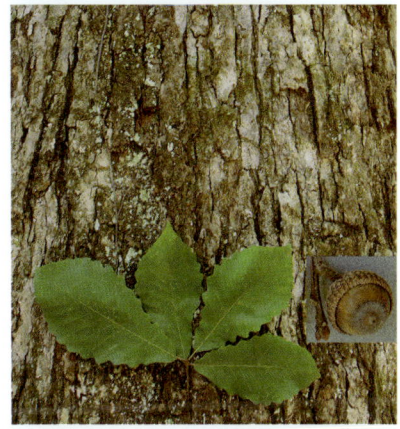

Swamp chestnut oak. Courtesy Adrian F. Van Dellen

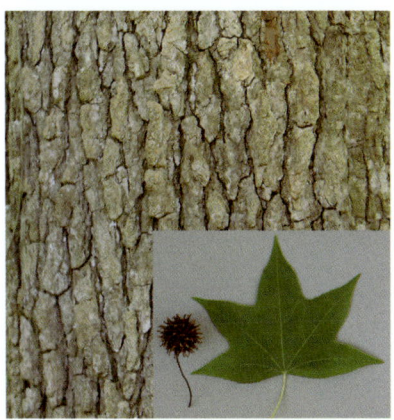

Sweetgum. Courtesy Adrian F. Van Dellen

Water Oak (*Quercus nigra*)

A native species, the water oak is a popular tree and has been widely planted in lawns, parks, and along roads. This large deciduous tree has small, variably shaped leaves (some appear to be spatula-shaped) and small acorns. The bark is tinged a light brown and is smooth.

Willow Oak (*Quercus phellos*)

The willow oak, compared to the water oak, has much narrower, willow-like leaves. The deciduous leaves are 2–5 inches long and ½–1 inch wide. These trees can reach 130 feet in height and 6 feet in diameter. Acorns from the willow oak are eaten by bluejays, wild turkeys, doves, rodents, squirrels, and the gray fox. Its wood is used for charcoal, shingles, fuel, and other general construction materials.

Flowers

Cardinal Flower (*Lobelia cardinalis*)

A showy perennial, the cardinal flower bears dozens of brilliant red flowers in summer that provide nectar for sulfur butterflies and hummingbirds.

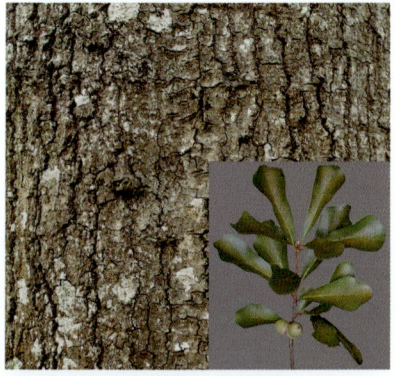

Water oak. Courtesy Adrian F. Van Dellen

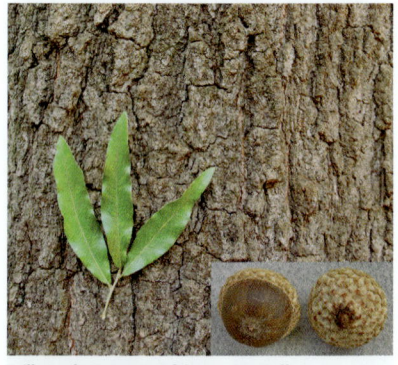

Willow oak. Courtesy Adrian F. Van Dellen

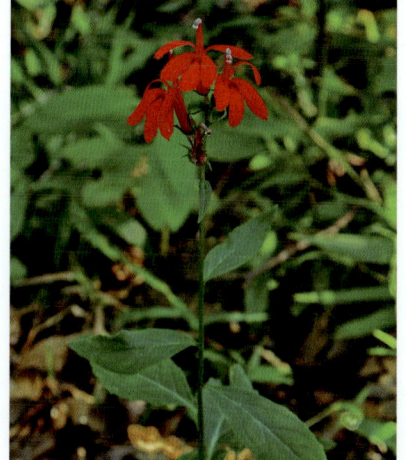

Cardinal flower. Courtesy Adrian F. Van Dellen

Neches River Rose Mallow (*Hibiscus dasycalyx*)

The Neches River rose mallow, a native hibiscus, is an endangered perennial that has delicate, slender, finely divided leaves and showy white flowers with dark, burgundy centers. These plants bloom from June through August but, depending on water availability, will sometimes bloom into October. The Neches River rose mallow is found only in East Texas along the Angelina, Neches, and Trinity rivers.

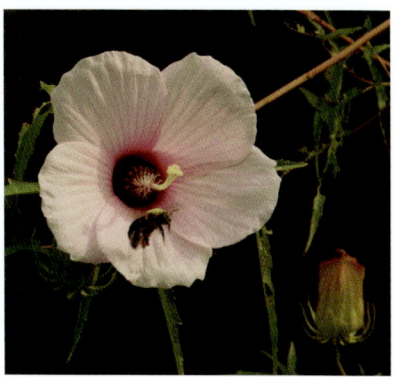

Neches River rose mallow. Courtesy Adrian F. Van Dellen

PROTECTING THE WILD NECHES

The accelerating loss of bottomland habitat, particularly forests with mature composition and structure, is of grave concern to many conservation groups as larger natural areas are needed to sustain numerous wildlife species, including millions of migratory songbirds and waterfowl.

Texas' remaining stands of bottomland hardwood forests are highly fragmented and threatened by residential and commercial development, agricultural conversion, timber removal, and infestation by invasive plants. A watershed-scale conservation effort is required to protect the Neches River's unique bottomland ecosystem, which sustains numerous wildlife species, including freshwater fish, mussels, amphibians and the "2,000 birds per square mile known to inhabit the Neches River bottomlands during winter" (Bob Parvin, *Bottomland Hardwoods, Every Acre Counts*, Texas Parks and Wildlife Magazine, October 1986).

The National Audubon Society recently released a forty-year bird census that stated habitat loss is the number one cause of declining bird populations throughout North America. The study revealed populations of our most common birds have dramatically plummeted; for example, the northern pintail, a spe-cies that utilizes the Neches River's riparian habitat, has declined 77 percent and the little blue heron, another inhabitant of the Neches, has declined 54 percent (National Audubon Society, State of the Birds, 2007).

Proposed reservoir projects would flood more than 175,000 acres of prime bottomland forest and drown over 140 miles of river. Water diversions would reduce downstream river flows, making the mixture of plant species in the Big Thicket National Preserve and other public lands less diverse. Clear-cutting forests and dividing large tracts into small holdings fragment wildlife habitat in the Neches Valley.

Tremendous local support for the Neches River National Wildlife Refuge led to its creation in Anderson and Cherokee counties in 2006. Biologists rate the bottomland hardwood forests of this site as among the best and least-disturbed in Texas. The U.S. Fish and Wildlife Service anticipates adding large tracts of land to the Neches Refuge over the next few years, with the refuge eventually totaling about 25,000 acres.

Congressman Kevin Brady is spear-heading a proposal to double the size of the Big Thicket National Preserve and greatly expand its potential to generate

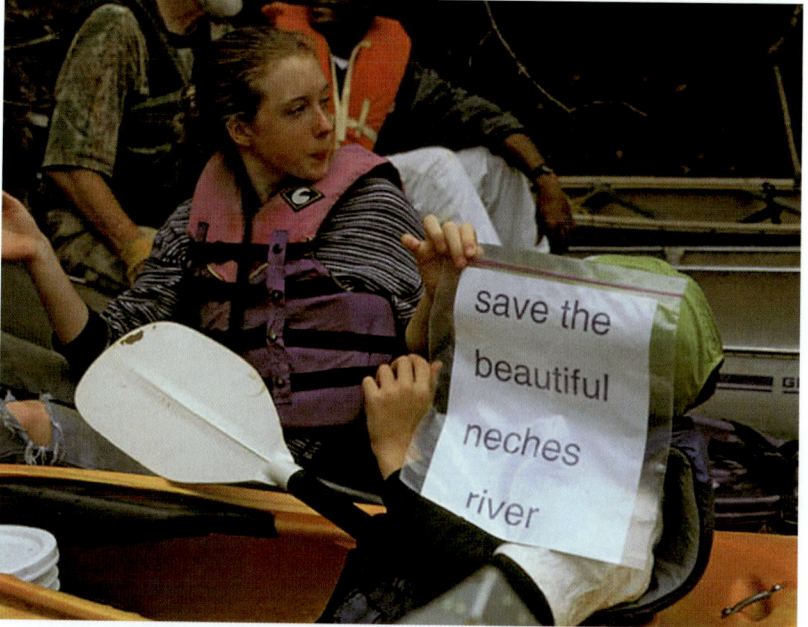

Paddlers advocating for saving the Neches. Courtesy Gina Donovan

tourism in the region. Internationally known for its biological diversity, the Big Thicket National Preserve is a jewel of the national park system and is designated as an international biosphere reserve. The proposed additions to the preserve—all to be acquired from willing sellers—would connect now-isolated tracts of land and protect the stream and river corridors of the Neches watershed. Construction of a canopy walk for visitors to view life in the treetops, an additional visitor center, and other interactive educational features would enhance the preserve's economic benefit to Texas.

A coalition of Texans, reflecting a broad spectrum of interests, support designating the Neches a national scenic river. Such a designation would protect the river from being dammed, while allowing other uses to continue as before. Rivers added to the Wild and Scenic Rivers System in other parts of the country have proved to be significant tourist draws, attracting paddlers, hikers, hunters, fishermen, and wildlife watchers from all over the country. Texas Conservation Alliance invites readers to learn more about the proposed Neches Scenic River at www.TCAtexas.org.

CONTRIBUTORS TO THE NECHES RIVER USER GUIDE

Texas Conservation Alliance

Texas Conservation Alliance, a forty-year-old statewide conservation organization, has led many successful projects to protect wildlife habitat in Texas. The alliance harnesses the energies and experience of Texans from varied backgrounds who share a common interest in protecting nature for our children's future. Texas Conservation Alliance builds grassroots coalitions of conservationists, sportsmen, landowners, advocates for nature tourism and outdoor recreation, business people, timber industry leaders, and elected officials to influence public policies and solve natural resource problems. This dynamic group of individuals and organizations has an exceptional record of protecting Texas' rivers, forests, coastlines, wildlife, and other natural habitats. The Texas Conservation Alliance and its representatives have been honored with the Theodore Roosevelt Conservation Award, Chevron Conservation Award, Sol Feinstone Environmental Award, and numerous other national, state, and local honors.

Houston Audubon Society

The Houston Audubon Society (HAS) is an autonomous chapter of the National Audubon Society, and is a 501(c)(3) organization that promotes the conservation and appreciation of birds and wildlife habitat. Incorporated in 1969, HAS is one of the most active chapters in the country and one of a very few that own and manage significant conservation property. HAS owns and manages seventeen sanctuaries (nearly 3,400 acres of natural habitat) two of which are internationally known. The High Island and Bolivar Flats Shorebird Sanctuaries are critical stopover, nesting, feeding, and roosting areas for migrating birds. More than 15,000 people from the United States and fifteen foreign countries visited these critical habitat areas in 2006. The organization has over 5,000 members in the Houston ten-county region and reaches over 25,000 children and adults annually through its conservation education programs.

The Conservation Fund

The Conservation Fund, formed in 1985, is a national nonprofit 501(c)(3) dedicated to balancing economic and environmental objectives to preserve America's land legacy—its natural, cultural, and historic heritage—for current and future generations. The Conservation Fund is a non-membership, non-advocacy organization. Through its partnership-driven approach, the fund works with corporations, governments, and communities to develop innovative solutions for Texas' complex conservation challenges. The fund has conserved more than 160,000 acres in Texas and over 5 million acres nationally. In 2007, it launched the Texas Pineywoods Experience—the Lone Star State's most ambitious land conservation and sustainable tourism initiative. The Pineywoods Experience is inventorying and documenting the natural, cultural, and historic riches of the region; marketing the Pineywoods as a public recreation destination; promoting the creation of new jobs based on this increased interest; and continuing a strategic and deliberate approach to land conservation, linking vital tracts of forest and parkland for people and wildlife.

USDA Forest Service

Established in 1905, the Forest Service is an agency of the U.S. Department of Agriculture. The Forest Service manages public lands in national forests and grasslands that encompass 193 million acres of land nationwide, which is an area equivalent to the size of Texas. The national forests and grasslands in Texas are made up of the Angelina, Davy Crockett, Sabine, and Sam Houston national forests located in the Pineywoods of East Texas, and the Caddo and Lyndon B. Johnson national grasslands located north of the Dallas–Fort Worth Metroplex. Altogether, the national forests and grasslands in Texas comprise 675,807 acres in fifteen counties. The Forest Supervisor headquarters for Texas is in Lufkin, and the agency employs about 174 full-time employees.

Stephen F. Austin State University Pineywoods Native Plant Center

The Stephen F. Austin State University Pineywoods Native Plant Center is a 40-acre green space sanctuary located in the heart of historic Nacogdoches, Texas. It was dedicated in 1999 with a mission to promote the education, conservation, and use of native plants of the southern forest. The property has 2 miles of handicap accessible trails winding through demonstration gardens and diverse habitats, including a bottomland hardwood forest.

Texas Parks & Wildlife Department

The mission of the Texas Parks and Wildlife Department (TPWD) is to manage and conserve the natural and cultural resources of Texas and to provide hunting, fishing, and other outdoor recreation opportunities for the use and enjoyment of present and future generations.

The TPWD's primary functions are the management and conservation of the state's natural and cultural resources, provision of outdoor recreational opportunities, conservation education and outreach, and cultural and historical interpretation. To this end, the TPWD operates and maintains a system of public lands, including state parks, historic sites, hatcheries, and wildlife management areas; monitors, conserves, and enhances the quality of public and private lands, rivers, streams, lakes, coastal marshes, bays, beaches, and gulf waters; manages and regulates fishing, hunting, and boating activities; assists public and private entities in providing outdoor recreational opportunities; conducts education and outreach events and programs; and cooperates with other governmental entities in this area.

Texas State University River Systems Institute

The Texas State University River Systems Institute demonstrates its deep commitment to the careful stewardship of the world's freshwater resources through its programs and projects. The institute affirms the unique role of water in our lives through collaborative research, public advocacy, and education on river systems. The institute is dedicated to preserving and protecting one of the earth's most remarkable resources, the irreplaceable gift—water.

T. L. L. Temple Foundation

The T. L. L. Temple Foundation was established in Lufkin in 1962 by members of the Temple Family to promote philanthropy, and enhance the quality of life for the inhabitants of the deep east texas pine timber belt—which is the geographical region where the family operated its business enterprises.

Recommended Reading

Abernathy, Francis E. *Tales from the Big Thicket*. Austin: University of Texas Press, 1974.

Ajilvsgi, Geyata. *Wildflowers of the Big Thicket, East Texas and Western Louisiana*. College Station: Texas A&M University Press, 1979.

Donovan, Richard M. *Paddling the Wild Neches*. College Station: Texas A&M University Press, 2006.

Fritz, Edward C. *Realms of Beauty: A Guide to the Wilderness Areas of East Texas*. Austin: University of Texas Press, 1986.

Gunter, Pete. *The Big Thicket: An Ecological Reevaluation*. Denton: University of North Texas Press, 1993.

Loughmiller, Campbell, and Lynn Loughmiller. *Big Thicket Legacy*. Austin: University of Texas Press, 1977.

Sitton, Thad. *Backwoodsmen: Stockmen and Hunters along a Big Thicket River Valley*. Norman: University of Oklahoma Press, 1995.

Sitton, Thad, and James H. Conrad. *Nameless Towns: Texas Sawmill Communities, 1880–1942*. Austin: University of Texas Press, 1998.

Sitton, Thad, and C. E. Hunt. *Big Thicket People*. Austin: University of Texas Press, 2008.

Stahl, Carmine, and Ria McElvaney. *Trees of Texas*. College Station: Texas A&M University Press, 2003.

Truett, Joe C., and Daniel W. Lay. *Land of Bears and Honey: A Natural History of East Texas*. Austin: University of Texas Press, 1984.

Watson, Geraldine Ellis. *Reflections on the Neches*. Denton: University of North Texas Press, 2003.